D0937971

THE SANDSTONE ARCHITECTURE OF THE LAKE SUPERIOR REGION

THE SANDSTONE ARCHITECTURE OF THE LAKE SUPERIOR REGION

KATHRYN BISHOP ECKERT

Wayne State University Press Detroit

Great Lakes Books

A complete listing of the books in this series can be found at the back of this volume.

Manufactured in the United States of America.
04 03 02 01 00 5 4 3 2 1

Library of Congress Cataloging-in-Publication Data

Eckert, Kathryn Bishop.
The sandstone architecture of the Lake Superior region / Kathryn Bishop Eckert.
p. cm.—(Great Lakes books)
Includes bibliographical references and index.
ISBN 0-8143-2807-5 (alk. paper)
1. Sandstone buildings—Lake Superior Region. 2. Regionalism in architecture—Lake Superior Region. 3. Sandstone—Lake Superior Region. I. Title II. Series.
NA722.E34 2000
721'.0441—dc21 99-36667

Contents

PREFACE

This book has its origins in my childhood friendships and in my career in historic preservation. A Detroiter by birth, I heard stories about Michigan's Upper Peninsula at an early age. The Sven Eklund family of Ishpeming, who were my family's neighbors in Ann Arbor when my father was in law school there, told me of their way of life in the Marquette Iron Range. The Richard Delbridge family were our neighbors in Detroit when I was in grade and high school, and Dr. Delbridge, who had been a second-generation Cornish-American miner in Hancock, and Mrs. Delbridge, an art teacher in the Detroit Public Schools, instilled in me a fascination for the Copper Country. Not only did they prepare pasties for us, they displayed in their house magenta and turquoise—the colors of Lake Superior sandstone and oxidized copper. Mrs. Delbridge painted the walls of their living room a soft dusty rose and laid a carpet of the same color—to enhance the iridescent turquoise glaze of the Pewabic tile of their fireplace.[1] Thus, when confronted with the physical presence of the Lake Superior region as an adult, I knew intuitively that the identity of the place was strongly tied to its native red sandstone buildings.

Soon after completing my master's degree in art history at Michigan State University in 1973, I began work on a historic sites publication at the then Michigan History Division (now the Michigan Historical Center). The historic preservation program, which began with the passage of the National Historic Preservation Act of 1966, was at that time only four years old in Michigan. It operated under the auspices of the division and the Department of Natural Resources. Soon I transferred into the division's historic preservation work.

To work effectively in historic preservation, I knew I must become more knowledgeable about American architecture. I entered the Ph.D. program in American Studies with emphasis on American literature, American history, and art history at Michigan State University. Seeking out the best people I could find with whom to study architectural history and popular culture, I acted on the advice of Visiting Professor Elizabeth Holt. I turned to Russel Nye in the English Department for work in popular culture,

Richard White in the History Department for readings on the American frontier, and to Leonard K. Eaton of the College of Architecture and Urban Planning at the University of Michigan for readings in American architectural history. I also traveled to Victoria, British Columbia, to take the 1976 Summer Institute in Cross-Cultural Studies with Alan Gowans to learn about social function in architecture. My dissertation, "The Sandstone Architecture of the Lake Superior Region," was completed in 1982. Revising the dissertation for publication was interrupted as I undertook the writing and editing of *Buildings of Michigan* (New York: Oxford University Press, 1993), the first volume in the Society of Architectural Historians' projected Buildings of the United States series.

After completing my graduate studies, I prepared a report, "The Sandstone Quarries of the Apostle Islands: A Historical Narrative," for the National Park Service, U.S. Department of the Interior, in 1985. The report was intended to provide interpretive information on one of the major resources of the lakeshore for use by staff in developing programs and media for visitors. I have also read papers about the quarries and buildings at professional meetings of the Society for Industrial Archaeology and the Society of Architectural Historians and have spoken about the subject to members of the Historical Society of Michigan and the Marquette County Historical Society.

The purpose of this book is to establish the importance of the sandstone architecture of the Lake Superior region to scholars and the general public. The architecture flourished from the early 1870s to the first decade of the twentieth century. The book relates the sandstone architecture to the quarries of the Jacobsville geological formation and the Bayfield group. The reddish brown sandstone architecture is in a profound sense representational of the region's cultural situation during this period.

This book places the sandstone architecture of the Lake Superior region in a total cultural perspective. It identifies and describes the native red sandstone architecture of the portion of the Lake Superior region of northern Michigan, Wisconsin, and Duluth, Minnesota, that lies within or immediately adjacent to the boundaries of the geological formations from which the red sandstone was extracted. It analyzes the historic and environmental forces that interacted to produce a distinctively regional architecture: land that contained copper, iron, and sandstone deposits; eastern and urban investors, who, speculating in land, resources, and industries, brought their attitudes and values to the region; immigrants who provided a labor force and brought their various native cultural attitudes and skills; and a rush of activity in copper and iron mining, shipping, and land speculation that created the need for all kinds of buildings.

The passage of time has not always been kind to the once sturdy sandstone buildings. The demolition of Hubbell Hall at Michigan Mining School, now Michigan Technological University, in Houghton in 1968, two years after the enactment of the National Historic Preservation Act and one year before the state historic preservation program began in Michigan, was a wake-up call to the people of the Copper Country. It sparked interest in protecting their remaining sandstone architecture. The loss of the locally beloved solid red sandstone J. H. Kaye Hall at Northern State Normal School, now Northern Michigan University, in Marquette in 1975 prompted the Marquette County Historical Society to promote quickly the listing of sandstone buildings in Marquette in the National Register of Historic Places. The Marquette City Hall was listed in the national register in 1975, and the designation of the Marquette Prison and other sandstone buildings soon followed.

Meanwhile, taking as a point of departure the broadbrush inventory of historic properties in the western Upper Peninsula undertaken in 1972 by Philip E. Metzger, a graduate student in the School of Natural Resources at the University of Michigan, I began to conduct intensive research on three of the most significant places Metzger had identified. My first historic preservation studies at the Michigan History Division were the national register nominations for the Calumet (Red Jacket) Downtown Historic District, the Calumet (Red Jacket) Fire Station, and the Calumet and Hecla Industrial District. All three nominations were approved for listing on the National Register of Historic Places in 1974. This assignment brought me north across the bridge and allowed me to experience the Upper Peninsula first hand. In rapid succession in the 1970s and 1980s, many of the historic buildings of the Upper Peninsula were identified and entered in the National Register of Historic Places and the State Register of Historic Sites. Almost always among the nominations were marvelous Jacobsville sandstone buildings. Serving on the Keweenaw National Historical Park Advisory Commission since the commission was seated in 1993, I have participated in the preparation of the General Management Plan and Environmental Impact Statement for the park. This continued involvement in the heritage and preservation of the Keweenaw has enabled me to help nurture and to witness the effort of people of the peninsula to protect their Jacobsville sandstone buildings.

The restoration and rehabilitation of many sandstone buildings in the Lake Superior region is both remarkable and heartening. In Marquette, Michigan, the Marquette County Board of Commissioners undertook the restoration of the Marquette County Courthouse; Peter and Barbara Kelly lovingly restored the Andrew and Laura Ripka House for their family to live in, and the Marquette County Savings Bank serves as professional offices for Dr. Kelly; parishioners of Saint Paul's Episcopal Church serve as stewards of their exquisite Gothic church; the Diocese of Marquette rehabilitated and maintains Saint Peter's Roman Catholic Cathedral; and the Detroit and Northern Bank has adapted the Charles Meeske House of the Upper Peninsula Brewing Company for new use. In Hancock, Michigan, the Hancock City Council has rehabilitated and added to the Hancock City Hall, and Suomi College has rehabilitated its rugged Old Main. In Calumet, Michigan, the village council maintains the Calumet Village Hall and Calumet Theater.

In Wisconsin, the Bayfield County Courthouse at Bayfield now serves as the headquarters for the Apostle Islands National Lakeshore, the Washburn State Bank at Washburn today houses the Washburn Historical Museum and Cultural Center, the Union Depot of the Wisconsin Central Railroad at Ashland is a restaurant, and the Ashland Post Office became Ashland City Hall. All of this work has been financed with both private funds and federal and state funds. Some private property owners rehabilitating income-producing properties have utilized the federal historic preservation tax credits. Other rehabilitations and restorations of sandstone buildings currently are underway in Calumet, Michigan, and in Duluth, Minnesota. Although these are some of the successes of rehabilitation and restoration efforts, other structures are at risk.

Many sandstone buildings have been lost to demolition and decay, and many sandstone quarries have been obliterated. Lost are the Alger County Courthouse at Munising, Michigan, the Peter White Hall of Science and Longyear Hall at Northern Michigan University in Marquette, the Richard P. Traverse House at Marquette, cellblock wings of the Marquette Prison, industrial buildings of the mining companies, and the Knight Block at Ashland, Wisconsin. Many quarry sites are overgrown or the victims of land use changes.

Even when people abandon or destroy their historic sandstone buildings in favor of new construction, their love of the sandstone building material persists. Although the city of Marquette discarded its historic sandstone and brick city hall, it used Marquette brown sandstone to veneer the exterior wall panels at the entrance to the new city hall that replaced the 1893 masterpiece.

I am fascinated with red sandstone architecture because of its association with the people who live in the Lake Superior region. Buildings tie people to place through their intended use, through their social function or symbolic function, and through their employment of indigenous material. The Marquette County Courthouse in Marquette, for example, serves as a suitable place in which to conduct the business of county government and as a visual metaphor of the institution of county government. The use of indigenous reddish brown Portage Entry sandstone trimmed with purplish brown Marquette raindrop sandstone from the Burt quarry ties the copper-domed courthouse to Marquette and the iron range, as does its intended use and its social function. This structure is a county icon. Assisted with funds from the Economic Development Administration, the citizens of Marquette County restored the grand old building in 1984. Local residents emotionally attached to the courthouse have ensured that the historic building will continue to serve as offices and courts of county government.

Extensive fieldwork over twenty-five years of investigating the quarry sites, now covered with one hundred years of overgrowth, and of inspecting the impressive stone buildings in all seasons of the year reinforced my belief in the unique characteristics of the red sandstone architecture of the Lake Superior region. Newspaper accounts identified and revealed the activities of many quarry companies, clients, architects, and contractors, and expressed the pride of local residents toward this native material and its role in building up their villages and cities. Minutes of meetings of boards of commissioners, and village and city councils, church vestries, and boards of governors; minutes of meetings of building committees; and correspondence between private clients and their architects made known the wishes, dreams, and fears of those imposing a civilization on this wilderness and demonstrated the role of native red sandstone in expressing these feelings. Professional geological accounts and interviews with geologists at the Michigan and Wisconsin Geological Surveys and regional universities clarified many of the scientific aspects of the stone and the formations, including the aerial extent of the formations, the physical properties of the stones and the similarities and differences between them, and the suitability of the stone as a building material. Company records and mineral statistics pinpointed the locations of quarry sites, identified investors, and offered some information on production. These research sources documented the emergence of the architecture of the Lake Superior region.

The current interest in regional midwestern architecture, vernacular architecture, cultural landscapes, and historic preservation makes the publication of this volume timely. For example, economists and planners interested in creating jobs in the Upper Peninsula have sought information on quarry sites to determine the feasibility of reopening several. In 1997–98 H. James Bourque, planner and economic developer, prepared a Hard and Soft Dimension Stone Feasibility Study under contract with the State of Michigan for the Michigan Jobs Commission on the feasibility of opening new quarries or reopening existing quarries in the Upper

Peninsula. The report identified five good sites from which stone might be extracted and examined the economics in terms of transportation, ease of access, quantity and consistency of the stone, infrastructure, equipment, ports, and the market. Proposed uses of the sandstone include replacement stone for rehabilitating existing sandstone structures and decorative and specialty products such as sills.

Sandstone architecture was built in places other than the Upper Peninsula, northern Wisconsin, and Duluth, Minnesota. Many red sandstone buildings stand in Minneapolis, Saint Paul, Milwaukee, Chicago, Detroit, Toronto, and other cities of the Upper Great Lakes. Sandstone for buildings also found its way to Buffalo, Saint Louis, Cincinnati, and New Orleans. Appendix 4, Representative Jacobsville and Bayfield Sandstone Buildings Elsewhere, will inform readers of sandstone's widespread usage.

I will not be satisfied until the sandstone architecture of the Lake Superior region is widely known among both the public and historians. Historians, architectural historians, cultural historians, and local residents need to examine the buildings with a fresh vision to fully comprehend the importance of sandstone to documenting and preserving the history of this region, and I hope this book will advance this cause.

ACKNOWLEDGMENTS

In writing this book, I was aided by many individuals and institutions in Michigan, Wisconsin, and Minnesota, and I want to acknowledge their help and cooperation. First I want to acknowledge the assistance of Leonard K. Eaton, Emil Lorch Professor Emeritus of the College of Architecture and Urban Planning, The University of Michigan, who directed my 1982 dissertation, "The Sandstone Architecture of the Lake Superior Region," and encouraged me to revise it for publication. At his suggestion I expanded the study from the sandstone architecture of the Upper Peninsula to include the buildings of Bayfield sandstone in northern Wisconsin and Duluth, Minnesota.

This book revises and updates my 1982 dissertation. Funding from the Graham Foundation for Advanced Studies in the Fine Arts enabled me to revisit sites and buildings along the south shore of Lake Superior to determine their present condition, their relation to their current surroundings and environment, and their ownership status. I conducted additional research and acquired photographs as well. I am indebted to the foundation.

I want to recognize the cooperation of numerous librarians and local historians. William Barkell of the Houghton County Historical Society, who, knowing of my project, sought out the J. W. Wyckoff papers and, who, with the other directors of the society, kindly made them available to me before they were transmitted to the Michigan Technological University Library Archives and Copper Country Collections. I want also to thank Erik Nordberg, and before him, Theresa Spense, of the Michigan Technological University Library Archives and Copper Country Historical Collections; Linda Paninen and Kay Hieble, and the late Esther Bystrom, before them, of the Marquette County Historical Society; Raymond Piiparinen and Esther Pekkala of Suomi College and the Finnish American Historical Archives; Marjorie F. Benton of the Bayfield County Historical Society; Alf A. Jentoff of the Baraga County Historical Society; James E. Lundsted of the Douglas County Historical Society; Ilene Schecter of the Library of Michigan; and Lynne Merrill-Francis of

American Institute of Architects of Michigan. Olaf Rankinen of Suomi College translated passages from the Finnish language.

I am grateful to Ford Peatross and Mary Ison of the Prints and Photographs Division of the Library of Congress for introducing me to the Detroit Publishing Company's extensive collection of photographs and assisting me with the Historic American Building Survey and Historic American Engineering Records. My thanks to Carolyn Pitts of the National Historic Landmark program, National Register, National Park Service, who made available National Register of Historic Places nomination forms for properties in Michigan, Wisconsin, and Minnesota. Kate Lidfors of the Apostle Islands National Lakeshore generously shared the agency's files on lakeshore sandstone quarries.

Several geologists furthered my understanding of the scientific aspects of the sandstone and the geological formations. Robert C. Reed of the Geological Survey Division of the Michigan Department of Natural Resources platted the aerial extent of the Jacobsville formation and pointed out similarities and differences between the Jacobsville, Freda, and Bayfield sandstones. Also cooperating were Harry O. Sorensen, Paul Daniels, Steve Wilson, and Milt Gere of the same division; Rod Cranson of Lansing Community College; M. G. Mundry of the Wisconsin Geological and Natural History Survey; and Richard W. Ojakangas of the Geology Department at the University of Minnesota-Duluth.

Historic preservationists working in the national register programs at the Michigan Historical Center and its predecessors of the Michigan Department of State and at the State Historical Society of Wisconsin aided my work. Also advancing my work were surveyors who identified historic sites in Michigan and Wisconsin. Surveyors who were especially helpful in my study were Philip E. Metzger; Ada Marlier and Patricia Williamson; Sue Sanborn and Lydia Wielenga; Kevin Harrington and Wendy Nichols, who recorded buildings in Calumet for the Historic American Building Survey in 1975; Charles K. Hyde, who identified industrial sites and engineering structures in the Upper Peninsula for the Historic American Engineering Record in 1977; and Linnie Thuma, who identified buildings and districts while working as an historic preservation planner for the Western Upper Peninsula Planning and Development Region in 1980 and 1981. I also want to acknowledge the special cooperation of Jeff Dean, State Historic Preservation Officer, who made available that office's survey data on the buildings of northern Wisconsin.

The readers for Wayne State University Press offered thoughtful comments and suggestions on the manuscript that greatly improved it, and I am indebted to them.

INTRODUCTION

It is difficult to think of the architecture of the Lake Superior region without thinking of the beautiful red sandstone that is indigenous to the region. Cropping out at various points along the south shore of Lake Superior, this rock is made up of sand-sized grains of quartz bonded together by iron oxide, calcite, authigenic quartz, and silica. It is the iron oxide that gives this lovely material its rich reddish brown color. These rocks comprise several geological formations—the Jacobsville formation and the Bayfield group of formations. They occur in a belt that underlies the land and the lake for nearly four hundred miles from Sault Sainte Marie, Michigan, to Duluth, Minnesota, and they crop out from Munising, Michigan, to the head of the lake.

Over the years nearly seventy companies extracted the famous sandstone from quarries in the Jacobsville formation and the Bayfield group. Scores of quarries along the lakeshore in Alger, Marquette, Baraga, Houghton, and Keweenaw Counties, Michigan, and in Ashland, Bayfield, and Douglas Counties, Wisconsin—but centered at Marquette and Portage Entry, Michigan, and Bayfield and Washburn, Wisconsin—provided stone that was easily and cheaply shipped by water to the region's most important commercial and industrial centers. The quarries were developed and operated by companies organized by investors residing in large midwestern cities, primarily Chicago, Detroit, and Milwaukee, and in the cities of the Lake Superior region, mostly Marquette, Houghton, Ashland, Bayfield, Superior, and Duluth.

A. P. Swineford, publisher of newspapers and mineral histories throughout northern Michigan, Wisconsin, and Minnesota, and then publisher of the *Marquette Mining Journal,* commented in 1872 on the prominence of the newly introduced building stone from beds and quarries at Marquette, L'Anse, Portage Entry, and Bayfield and its economic importance to the region: "In fact, all along the south shore of the lake are situated beds of this brown stone of the very finest quality, which in years to come will

Milwaukee County Courthouse, 1870–73, Leonard A. Schmidtner, Milwaukee, photograph 1900. Destroyed 1976. (Courtesy Library of Congress, Detroit Publishing Company Photographic Collection)

be developed and made to yield the material for the building up, in a more substantial manner, our great lakes cities. Indeed, such is the acknowledged value of our Lake Superior brown sandstone as a building material, that we may confidently expect our quarries, in a few years, to become one of the most important elements of substantial wealth."[1]

A later observer of the sandstone industry in the Upper Peninsula writing in *Michigan and its Resources* noted optimistically the economic value of stone as a resource to be exploited in the name of progress. "The upper peninsula is one of nature's great banking houses, whose capital stock is its deposits, subject to be drawn only by the pick and shovel; with no exacting cashier to count, no identification required, no security asked; simply locate the funds; open the doors; help yourself."[2]

The first major civic building to use Lake Superior sandstone was the Milwaukee County Courthouse. The Milwaukee County Board of Supervisors decided in the late 1860s to build a new courthouse to replace the county's plain wooden Greek Revival pioneer courthouse built in 1843 and enlarged in 1846. The new courthouse would stand on the northern portion of the Cathedral Square facing south on Juneau Park. Not only did the people of Milwaukee County need a new courthouse, they wanted one whose appearance and quality would honor them all.

In July 1868 the Committee on Public Buildings reviewed plans for the new courthouse prepared by Leonard A. Schmidtner, a Polish-American Milwaukee architect. Plans showed a large Renaissance Revival structure, approximately 210 feet by 130 feet in plan, executed in pressed brick with stone ornamentation. Two-story wings flanked the domed three-story central core, and giant porticos supported by Corinthian columns atop single-story Doric porticos adorned all four sides of the building. The interior was arranged with corridors thirteen feet in width intersecting a rotunda thirty-six feet in diameter. Offices were off the corridors. All corridors, stairways, and vestibules, as well as the rotunda, were painted with frescos.

However, the supervisors felt that public opinion favored constructing the grand public building in stone. Schmidtner claimed that facing the entire building in stone, whether Illinois white stone or Lake Superior brownstone, would cost little more than pressed brick with stone trim—and Lake Superior brownstone would be cheaper than Illinois white stone. But people hesitated using an untried stone in an important public building, and the engineers and contractors contemplating the construction of the courthouse questioned the use of Basswood Island brownstone.

So in 1869 the Milwaukee County Board of Supervisors sought the opinion of recognized scientists on the quality of the stone as a building material. A special committee consulted scientists Increase Allen Lapham and James Hall, respected Marquette resident Peter White, and other experts. Lapham (1811–76), a geologist and civil engineer, examined the stone and reported to the building committee as follows:

> The Bass Island (Lake Superior) sandstone having now been fully tested as to its power to resist crushing, its porosity, and as to its chemical composition, I am able to give a more correct and reliable opinion as to its suitableness for the facing of the walls of the new court house than that expressed in my letter to you of January 9th last.
>
> The strength of the stone is found to be such as to require a pressure of 5,426 pounds per square inch to crush it—a strength much more than is required for the purpose of our court house. The test was made under the direction of Prof. Joseph Henry, Secretary of the Smithsonian Institute, at the Navy Yard, in Washington. . . .

> In view of all the facts in the case, I cannot but recommend that the contract contemplating the use of the Lake Superior sandstone be carried out, care being taken to select the best of the stone, and to see that they are properly . . . placed in the wall.

Hall (1811–98), a geologist and paleontologist who had studied natural science with Amos Eaton at Rensselaer Polytechnic Institute, examined a sample of Basswood Island sandstone from the Bayfield group and reported, "The rock is a reddish brown sandstone of uniform texture and free from clay seams. For the purpose of building or for ashlar facing, I regard it as a valuable and durable stone. The character of the weathered surface and the presence of lichens show slow disintegration in its natural exposures."[3]

In September 1868 the Committee on Public Buildings had recommended to the board of supervisors the use of Lake Superior sandstone for the new courthouse. This opinion was echoed in a special petition signed by prominent citizens and recorded in the September 1868 Proceedings of the Milwaukee County Board of Supervisors. With statements from experts employed "to examine the quality of the stone" and "to report whether it is a proper and fit material" to use in the construction of the courthouse, the Milwaukee County Board of Supervisors concluded that Milwaukee brick veneered with purplish reddish brown Bass Island sandstone would be the "cheapest, most durable, and most attractive" material.[4]

Alanson Sweet (1804–91) and other Milwaukee men had prospected the Chequamegon Bay area for fifteen months to locate quality building stone for the proposed Milwaukee County Courthouse. Eventually, they selected a quarry site on Basswood Island, then Bass Island, and, in 1868, Sweet and Daniel Wells acquired their interest in the Bass Island Brownstone Company quarry. Sweet was a stone mason, politician, builder, and promoter of railroads and plank roads. After the Milwaukee County Board of Supervisors had approved the use of Basswood Island brownstone in the courthouse, Milwaukee and Chicago investors organized a company to extract the stone and ship it to Milwaukee.[5]

In the spring of 1870, Strong, French and Company began quarry operations on Basswood Island. The company continued extracting stone until October 1873, when financial panic and a lawsuit affecting portions of the title to the land and equipment closed down their operation. In its first season, the company employed approximately forty workers, including twenty-five that fall, and ten to fifteen that winter. Workers cleared timber and stripped the surface, installed machinery, and built docks. The stone was removed manually the first season. In spite of the work required to prepare the site, two thousand tons of stone were shipped to Milwaukee. The exclusive shipment of sandstone to Milwaukee to com-

Wisconsin scientist Increase A. Lapham (1811–76) is credited with furthering the use of Lake Superior sandstone for building purposes when he supported its use in the construction of the Milwaukee County Courthouse in 1868. From History of Milwaukee, Wisconsin *(Chicago, Western Historical Co., 1881).*

plete the order for the Milwaukee County Courthouse continued until mid-August 1871.

The cornerstone for the Milwaukee County Courthouse was laid on 7 September 1870, and the structure was completed in 1873 at a cost of $650,000. From the day it opened, the Milwaukee County Courthouse was declared a "Wisconsin Building" because its massive walls were of Lake Superior red sandstone from Basswood Island backed by pale yellow

Milwaukee brick.[6] The county abandoned this sandstone structure for a new Beaux-Arts Classical courthouse in 1930–31, and the old courthouse was razed in 1976.

Profit-motivated quarry investors and shareholders, stone yard owners, and businessmen in Marquette, Houghton, Ashland, Bayfield, and Duluth promoted the use of sandstone locally. Quarrymen in Marquette opposed the tendency of local builders to import brick at great expense or to use lumber in great quantities. The campaign they mounted against the use of brick and lumber in Marquette typified strategies employed in the Copper Country and Chequamegon Bay areas. The following article, which appeared in the *Marquette Mining Journal* for 17 August 1872, just three weeks before the incorporation of the Marquette Brownstone Company, summed up the arguments and conclusions expressed in their promotion:

> Nature has placed at our doors a material for building which far surpasses the brick or lumber, seemingly as if not to neglect a single duty which she owes so intelligent and enterprising a people. This material is the sandstone of which the new Methodist Church and the "Superior Buildings" are being erected. Strangers who notice the variegated color in the formation of this stone, and examine its texture, are heard to pronounce it as fine building material as they have ever seen, and ask, at once, why it has not long ere this been generally used; in addition to this variegated strata, all of which is of as good quality for building as any sandstone known, it is but a question of time when the major portion of the best buildings of Marquette will be reared by means of these blocks.

Sandstone from the quarries in the Jacobsville formation and the Bayfield group was shipped throughout much of the United States. It went to the Great Lakes ports of Duluth, Milwaukee, Chicago, Detroit, Buffalo, and Toronto and inland to such points as Saint Paul, Minneapolis, Kansas City, Saint Louis, Philadelphia, and New York City. In these places and elsewhere, Lake Superior sandstone furnished the building material for hundreds of buildings—buildings like the Tribune Building (1872) in Chicago, the Germania Bank Building (1888–90) in Saint Paul, the Cincinnati City Hall (1888–93), the Chamber of Commerce (1894–95) in Detroit, and the Waldorf-Astoria Hotel in New York City.

A look at three of these buildings reveals some of the reasons the appeal of sandstone from the Jacobsville formation and the Bayfield group extended to builders outside the area of the geological formations. For the builders of the Tribune Building, the sandstone's fireproof qualities set this material apart from other materials. After Strong, French and Company completed its order of Basswood Island brownstone for the Milwaukee

Tribune Building, 1872, Burling and Adler, Chicago, photograph date unknown. Destroyed 1902. After the Great Fire, Chicago builders sought durability above all in materials. (Courtesy Tribune Archives, McCormick Research Center, Wheaton IL)

County Courthouse in mid-1871, it began to direct regular shipments of the material to Chicago. Thus, on 8 October 1871, the eve of the Great Fire that swept through Chicago laying ruin to eighteen thousand buildings, rendering one hundred thousand people homeless and three hundred dead, and destroying more than two million dollars in property, Chicago stone yards held good supplies of Basswood Island stone. To continue providing stone to the Chicago market, quarriers Strong, French and Company operated through the winter of 1872–73. In the wake of the fire, Chicago builders sought durability above all else in materials, and the Basswood Island sandstone had won its acceptance in the Milwaukee County Courthouse. Moreover, its reddish brown color appealed to the emerging taste for highly colored, durable stone.

The Chicago fire destroyed the first Tribune Company Building, a brand-new four-story structure built on the southeast corner of Madison

Cincinnati City Hall, 1888–93, Samuel Hannaford, Cincinnati, photograph date unknown. (Courtesy Cincinnati Historical Society)

and Dearborn Streets of cast iron and Niagara limestone dug out of the Illinois and Michigan Canal. The Tribune Company immediately took steps to build a more elegant and convenient structure on the same site. To plan the new building, the company called on Edward Burling (1819–1902), the designer of the fire-ravaged structure. Burling, one of the first professional architects to practice in Chicago, was then associated in partnership with Dankmar Adler (1844–1900). Burling and Adler created a simple five-story commercial block and ordered it constructed of durable, dressed Lake Superior red sandstone quarried on Basswood

Detroit Chamber of Commerce, 1894, Spier and Rohns, Detroit, photograph date unknown. The building rises above a lower story base of Bayfield brownstone. (Courtesy Burton Historical Collection of the Detroit Public Library)

Island. On 9 October 1872, just one year after the fire, the Tribune Company published its newspaper in the new $250,000 building.

Great fires that razed buildings, blocks, and districts in Marquette in 1868, in Hancock in 1869, in Red Jacket in 1870, in Washburn in 1888, and in Bayfield in 1883, after these communities had built up and become more populous, inspired village and city councils to adopt fire codes and regulations and to acquire adequate firefighting apparatus and skills.

Marquette County Savings Bank, 1891–92, Barber and Barber, Marquette, photograph c. 1985. Carving detail from front exterior. (Courtesy Balthazar Korab)

Following these disasters most places enacted ordinances that required all structures within certain prescribed boundaries in the central portions of the communities to be constructed of brick or sandstone. The need to build with fireproof masonry greatly increased the demand for sandstone.

The Cincinnati City Hall exemplifies the use of reddish brown sandstone in combination with other stones in an important midwestern civic building. In 1888 the Cincinnati City Board of Supervisors commissioned Samuel Hannaford (1835–1910), a capable and prolific Cincinnati architect, to design its city hall. The result was a massive towered and turreted Richardsonian Romanesque structure built of Missouri granite, light grayish yellowish brown Amherst, Ohio, sandstone, and elaborately carved Bayfield brownstone.

The brownstone was quarried either by the Ashland Brown Stone Company at Presque Isle, now Stockton Island, or by the Prentice Brownstone Company at Houghton Point on the mainland in Bayfield County. The records are unclear as to who furnished the Bayfield stone for the Cincinnati City Hall: The *Bayfield Press* for 14 April 1888 reported that the Ashland Brown Stone Company had the contract for stone for a new city hall in Cincinnati but noted on 28 July 1888 that the Prentice brownstone quarry at Houghton Point had a contract to furnish "5 carloads of stone for City Hall at Cincinnati"; the Prentice Brownstone Company claimed in its promotional literature to have furnished the stone and supplied statements from the architect and city officials on their satisfaction with its durability, elegance, and beauty. There is no quarrel, however, that the Bayfield group in Wisconsin yielded the Lake Superior brownstone. The booklet entitled "Dedication of the New City Hall" (1893) states only that Wisconsin furnished the brownstone, but most likely the stone came from the Prentice Brownstone Company quarries at Houghton Point. The Cincinnati City Hall seemed to the people of Cincinnati the most magnificent city government building in the country. It cost $1.5 million.

The Detroit Chamber of Commerce Building further demonstrates the preference for red sandstone, in this case, as trim for a brick structure. The desire of Frederick H. Spier (1855–?) and William C. Rohns to work with stone, the familiarity of the Chamber of Commerce officials with the attributes of Lake Superior sandstone, and the promotional efforts of the Excelsior Brownstone Company in Detroit influenced the decision to clad the first stories of the Detroit Chamber of Commerce building with Bayfield sandstone from Hermit Island. Russell A. Alger, David Whitney, Thomas Palmer, and other Detroiters had speculated in the timber and mineral resources of the Lake Superior region, and they preferred red Lake Superior sandstone for so fine and important a building as the Chamber of Commerce that would represent trade and commerce in Detroit.

From 1870 until 1910, the prosperity of the copper and iron mining, lumbering, and shipping industries of the Lake Superior region called for ever more substantial buildings, and in satisfying this demand, architects, builders, and clients alike preferred the sedimentary rock from the Jacobsville formation and the Bayfield group. They found the stone suitable because it was beautiful, durable, and carvable. Moreover, because it was extracted easily in large blocks and shipped cheaply by water, it was economical. It was promoted aggressively. The red sandstone city halls, county courthouses, churches, schools, libraries, banks, commercial blocks, and houses they built give the Lake Superior region a distinct identity.

With solid and substantial red sandstone buildings bankers solicited and held depositors, landlords secured tenants for commercial space, mining companies demonstrated their paternalism toward employees threatening to strike, religious groups declared the values of their institutions and the social position of their members, educators called attention to the

Marquette City Hall, 1893–94, Lovejoy and Demar, Marquette, photograph c. 1985. (Courtesy Balthazar Korab)

importance of scholarship and the arrival of cultural maturity, governments denied economic decline, ethnic groups upheld the validity of their native traditions and institutions in the face of the melting pot, and citizens created a sense of place. Sandstone architecture symbolized permanence, solidity, and belonging and lifted the spirits of those who beheld and used it.

The Richardsonian Romanesque mode of architecture arrived in the region between 1880 and 1900. This style, which had already permeated most of the upper Middle West, brought exciting new potential to the architecture of the Lake Superior region. The mode was named for Henry Hobson Richardson (1838–86), whose designs for buildings borrowed motifs from the medieval French Romanesque and employed a picturesque massing of towers, porches, arches, and buttresses in varied colors and textures of stone. Not only was it a style of national prominence, but it was one that was particularly suitable to the character of the region. Its dynamic stone masonry, massive forms, and irregular outlines introduced qualities of style that almost seemed to beg for rendition in Lake Superior sandstone. Indeed, Richardson's own rock-faced style was intentionally geological in some of its imagery and thus potentially expressive of the physical structure of the Lake Superior region. The Richardsonian mode arrived in the Lake Superior region at the height of the lumbering and mining years and was received with flamboyant, if sometimes innocent enthusiasm. It appeared in buildings of all types but was used with the greatest originality in domestic architecture, especially in the cities and villages of the mineral ranges.

Art historian Alan Gowans has stated that the heavy stonework and ponderous forms of the Richardsonian Romanesque created an impression of solidity, an attitude of stability, and a mood of security that met the needs of Americans living in an era of paradox. On the one hand, Americans felt confident and exuberant in the midst of rapid change; on the other hand, they felt disturbed and afraid of the effects of change, upheaval, destruction, and demoralization. Solid, stable, and secure stone buildings symbolized and expressed a sense of permanence and belonging in an environment of uncertainty and impermanence.[7]

The Richardsonian mode marked the triumphant climax of nineteenth-century Romanticism in both American and Lake Superior region architecture. It was its final affirmation. The World's Columbian Exposition of 1893 in Chicago introduced a wholly different concept of formal relationships, a concept that within a decade would radically alter the direction of architectural design in this country. Its source was the Ecole des Beaux-Arts in Paris. In contrast to the aggressive picturesqueness of the Richardsonian mode, the design principles taught at the Ecole were modeled after the monumental architecture of ancient Rome and were classical in the highest degree; they espoused symmetry, regularity, harmony, and axial planning. The Ecole began attracting young Americans shortly after midcentury, and by 1890 a sufficient number had been trained there, or had otherwise been influenced by its teachings, to form a mature and outspoken group that was actively practicing its principles. It was these architects, under the leadership of McKim, Mead and White of New York, who were responsible for the design of the Chicago Fair. The grandiose plan that they produced was Beaux-Arts Classicism at its monumental best. It not only rivaled Rome in its colonnaded facades and vaulted interiors, but to create the impression of marble, the preferred material of classicism, most of the buildings were plastered and painted white.

Both the high classicism and the whiteness of the Columbian Exposition would affect the architecture of the Lake Superior region as quickly as it would the rest of the nation. Public architecture in particular was soon conceived in these classical terms. As a result, the red and brown local sandstone that had so enriched the rough polychrome walls of the Richardsonian style gave way to the smooth surfaces of marble and light gray and white limestones and light grayish yellowish brown sandstones that were mandated by classical design. This change in style was especially dramatic in the Lake Superior region, where local sandstone had been so conspicuously displayed.

This radical and swift change in architectural direction dealt a fatal blow to the red sandstone quarries of the Lake Superior region. Indeed, as the twentieth century progressed, stone in general gave way to artificial stone, concrete, and brick. To speed this process, brick manufacturers mounted a successful campaign against stone. Sandstone relinquished its preeminence to brick in the Lake Superior region in the 1890s and 1900s. William H. Jordy has noted that brick manufacturers flocked to Chicago in great numbers after the fire and introduced the technology of hard Philadelphia pressed brick, and brick construction rose in popularity there in the 1880s.[8] The effects of this campaign are recorded clearly in *The Brickbuilder,* "An Illustrated Monthly Devoted to the Advancement of Architecture in Materials of Clay," published in Boston beginning in 1892. The effort contributed to the falling from favor of sandstone architecture throughout the country and to the demise of the Lake Superior quarry industry. Brick manufacturers touted the desirability, beauty, and freshness of enameled brick and its ability to produce "beautiful designs in architecture and harmonious blending of colors."[9] Then, too, the rapid development of steel and reinforced concrete skeletal construction in Chicago in the mid-1880s encouraged the use of brick and other synthetic materials, which were lighter and more economical as sheathing for the frame.

The impact of the Columbian Exposition brought to an abrupt end one of the most expressively regional episodes in the Lake Superior region's architecture. Although local architects continued to design with a sensitivity for their time and place and outstanding geniuses continued to make their boldly individual statements, the architecture of the region became increasingly absorbed into the mainstreams of national development.

Yet building with red sandstone continued briefly and sporadically in the region well into the twentieth century. J. H. Kaye Hall (1913–15, Charlton and Kuenzli) at Northern State Normal School, now Northern

Michigan University, is a case in point. In several instances stone salvaged from earlier buildings was laid in the exterior walls of twentieth-century structures. Examples are the Baraga School (1903–1906, John D. Chubb), built of Marquette brownstone salvaged from the Grace Furnace in Marquette, and even the gas station on Third Street in Laurium. A more whimsical example is the replica of the *U.S.S. Kearsarge,* built of Jacobsville sandstone, and mine rock on U.S. 41 and Michigan 26 at Wolverine during the Works Progress Administration.

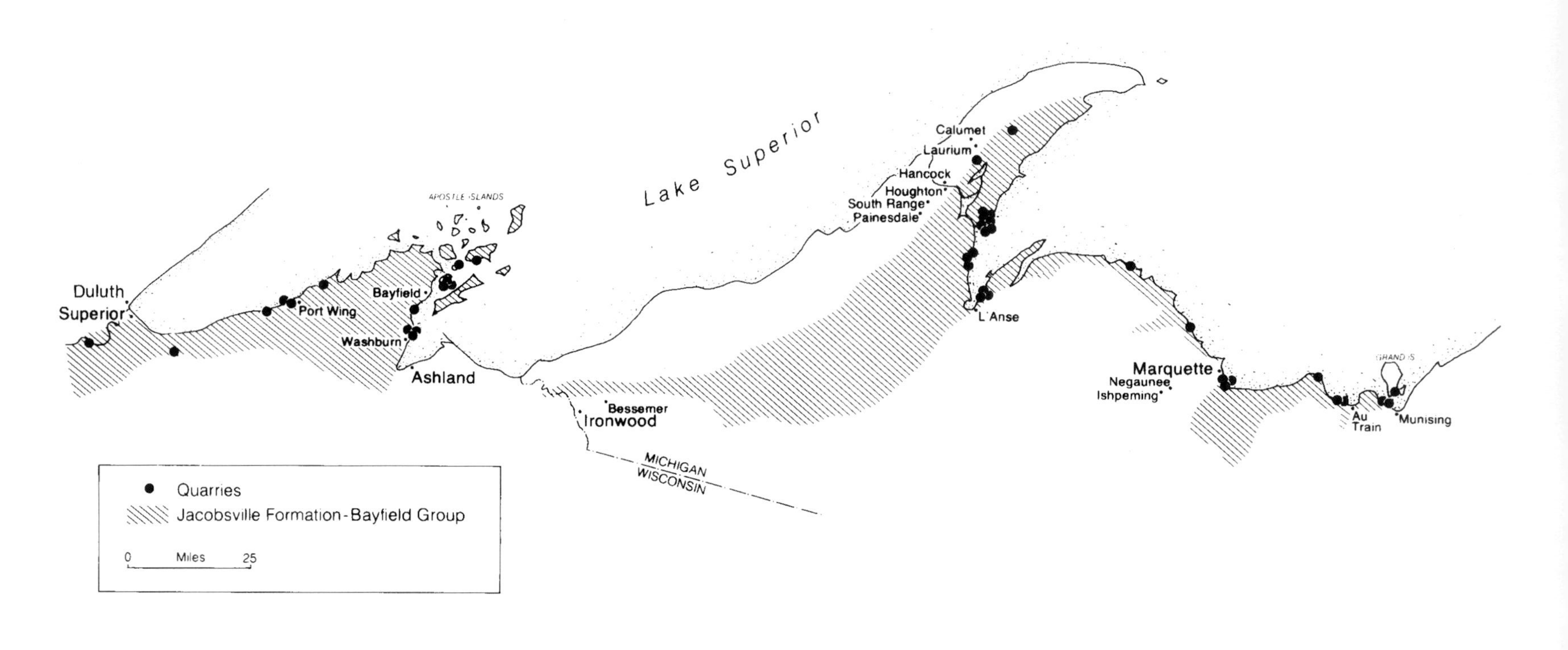

Map showing the distribution of quarries in the Jacobsville formation and Bayfield group. (Courtesy Sherman K. Hollander)

1

LAKE SUPERIOR SANDSTONE: THE JACOBSVILLE FORMATION AND THE BAYFIELD GROUP

INTRODUCTION

Reddish brown sandstones crop out along the southern shore of Lake Superior from Munising, Michigan, to Duluth, Minnesota, comprising the Jacobsville formation and the Bayfield group of three formations with similar properties. These are the Orienta, the Devil's Island, and the Chequamegon formations that together are known as the Bayfield group.

The most prolific outcropping, the Jacobsville formation, occurs on the Keweenaw Peninsula, southeast of the Keweenaw fault. It extends eastward along the shore of Lake Superior to Sault Sainte Marie and Sugar Island and probably constitutes the bedrock over much of the bottom of Lake Superior in that area.[1] From Munising to Beaver Lake, the Jacobsville formation is completely below water and is overlaid by the Munising formation, a light grayish white sandstone. Jacobsville sandstone is named after the village of Jacobsville at the entry to the Portage River on the Keweenaw Peninsula, some fourteen miles southeast of Houghton, where the red sandstone was quarried. Jacobsville, the village, was named after John H. Jacobs (1847–1934), a pioneer in the development of the Lake Superior sandstone industry.

The Bayfield group of three formations occurs in northern Wisconsin and extends westward from the Montreal River or, more precisely, from the mouth of Fish Creek at the head of Chequamegon Bay, including the Apostle Islands; to the Saint Louis River, which forms the border between Superior, Wisconsin, and Duluth, Minnesota; then southwestward into Minnesota. The sandstones of the Bayfield group underlie the lower ground along the entire lakeshore in Wisconsin and form the basement rock of all the Apostle Islands. The exposures of Lake Superior sandstone are restricted almost entirely to the shores of the lake on the mainland and the Apostle Islands. The Bayfield group is named after the village and the county on the mainland opposite the Apostle Islands, twenty miles north of Ashland. Bayfield village and county were named after Captain Henry Bayfield (1795–1855) of the British Royal Topographic Engineers, who

Michigan-Lake Superior Hydroelectric Power Company Plant (Edison Sault Power Plant), 1896–1902, Hans A. E. von Schon, engineer; James Calloway Teague, Sault Sainte Marie, photograph 1902. Sandstone excavated from the earth and discarded as the canal was constructed rises in the walls of the steel-frame structure. (Courtesy Edison Sault Electric Company)

first explored the sandstone of the Lake Superior region while conducting his trigonometrical survey of the Lake Superior region in 1822–23.

The Chequamegon formation underlies all of the Apostle Islands southeast of a line from Sand Point to Devil's Island and is exposed at Houghton Point and on the shore cliffs south of Washburn. The Devil's Island formation, named for the island that displays exposures of the formation, underlies the Chequamegon formation. The Orienta formation, named for the quarry town on the Iron River thirty-five miles east of Superior, underlies the Devil's Island sandstone and the Western Plain from the line of outcrop of the Devil's Island formation all along the south shore of Lake Superior north of the Douglas Range and into Minnesota.

Until the early twentieth century, the reddish brown sandstone of the Lake Superior region was generally and popularly called Lake Superior sandstone, brownstone, or redstone and prefixed specifically by the name of the place in which it was quarried, such as Marquette, Portage Entry, or Bayfield sandstone.[2]

Geological Aspects

The sandstone of the Jacobsville formation was formed after early Cambrian streams flowing northward deposited sand and gravel sediment along the lowlands that lay along the present lakeshore at the northern edge of the Upper Peninsula of Michigan and at the northern edge of Wisconsin. Streams flowing northward from the Northern Michigan Highlands, remnants of mountains initially formed during an episode of igneous intrusion and crustal deformation known as the Precambrian Penokean Orogeny, and extending in a belt from Ontario on the east across the central and southern part of the Upper Peninsula and into Wisconsin, carried sand and gravel down to the lowlands that lay along the south shore of Lake Superior. On reaching gentler lowland slopes, the early Cambrian streams slowed down and dropped their load of sediment in shifting channelways or in local lakes. The coarse sand and gravel deposits and interbedded shales that resulted included fragments of many rocks and minerals eroded from preexisting rocks of the source area in the Northern Michigan Highlands. These deposits formed the sandstone of the Jacobsville formation. Because of the highly irregular Precambrian surface on which it was deposited, the Jacobsville formation varies from 46 to 1,100 feet in thickness.[3] The thickness of the Bayfield group is estimated to be 4,300 feet in the Lake Superior syncline.

The Jacobsville formation and the Bayfield group appear to be related to each other. Heavy mineral suites, bulk mineralogy, sedimentary features, and paleomagnetic measurements demonstrate their great similarity. They are probably of late Keweenawan in age, that is, more than six hundred million years old.[4]

The physical properties and uses of sandstones depend on the mineralogy of the sand grains and the material that cements them together. Common bonding agents are clays, iron oxides, calcite, and silica. The type of cement and the degree of cementing influences the durability, strength, and, ultimately, the use of sandstones in building.[5] The sandstones of the Jacobsville formation and the Bayfield group are quartz sandstones cemented with iron oxide, calcite, and, in the case of the former, authigenic quartz, and in the case of the latter, silica.[6] Their physical appearance and properties are similar. Typically, sandstone is composed of grains, one-sixteenth to two millimeters in size, that are bonded together to form the rock. The grains of the sandstones of the Jacobsville formation and Bayfield group are one-fourth to one-half millimeter in size but may range from shale to conglomerate. The small grains probably were responsible for the fine-grained appearance of the top grade sandstone quarried in the Portage Entry vicinity, which was said to have a "liver texture."

The major constituents of Jacobsville sandstone based on an average of samples collected in Marquette and Alger Counties are nonundulatory quartz (27.4 percent) and undulatory quartz (27.0 percent). Also present are potassium feldspar (23.0 percent), silicic volcanic clasts (12.3 percent), polycrystalline quartz (3.8 percent), metamorphic (2.4 percent), sedimentary (1.4 percent), opaques (1.3 percent), onafic volcanic (0.8 percent), and plagioclase (0.1 percent).[7]

The major constituents of Orienta sandstone, one of three formations within the Bayfield group, vary regionally, but based on an average of fifty-two samples are nonundulatory quartz (33.3 percent) and undulatory quartz (29.7 percent). Also present are potassium feldspar (17.3 percent), silicic volcanic clasts (9.4 percent), mafic volcanic fragments (1.6 percent), polycrystalline quartz (3.9 percent), opaques (2.3 percent), metamorphic (0.9 percent), sedimentary (0.7 percent), and plagioclase (0.4 percent).[8]

The heavy mineral suite of the Jacobsville sandstone is ilmenite (84 percent), leucoxene (4–5 percent), garnet (4 percent), apatite (3 percent), zircon (3 percent), and tourmaline (1 percent). The heavy mineral suite of the Orienta sandstone of the Bayfield group is ilmenite (78 percent), leucoxene (13 percent), apatite (3–4 percent), zircon (3 percent), garnet (2 percent), and tourmaline (1 percent).

Small quantities of iron oxides give the rocks of the Jacobsville formation and the Bayfield group their striking red color. Red and reddish brown predominate, but the color ranges from strong red to brownish red to purplish red, and the basic red may be mottled with white streaks, blotches, and circular spots. Spherical light spots caused by leaching and bleaching are common features of the sandstones of the Jacobsville formation and the Bayfield group. The boundary between the red and white colors is sharp and well defined, showing little or no gradation. Generally, shale beds and fine-grained sandstone units possess the greatest intensity of red coloration and the least amount of white mottling. Massive coarse-grained units are white or light pink mottled with red streaks. Leaching of the red color follows planes of cross-bedding and produces alternating red and white streaks parallel to the stratification.[9] The massive sandstone facies—the part of a formation that is distinguished by massive and relatively persistent bedding—in most outcrops of the Jacobsville formation and the Bayfield group are light red, reddish brown, purplish reddish brown, purplish pinkish brown, pink, or white. Solid colors predominate, but the conspicuous white mottling so striking in the lenticular sandstone facies is lacking in many exposures. Some beds are dark purple with minor mottling.

Historical Aspects

The historical facts that lie behind the development of the use of stone for building purposes are themselves revealing. In the early nineteenth century, before the quarries were opened, scientific surveys sponsored by federal and state governments provided increasingly accurate and reliable data about mineralogical resources, floral and fauna, and soil conditions. The surveys noted the reddish brown Lake Superior sandstone. The resulting reports prepared by these geologists and land surveyors described the physical characteristics of the sandstones and gave their location. They also forecasted their marketability and importance to the economic development of the state and the region, alluding to their usefulness as a building material in building up local cities and villages.

As early as 1819, exploratory expeditions under the authority of the U.S. Secretary of War, together with Lewis Cass (1782–1866), the territorial governor of Michigan, were carried out by scientist and writer Henry Rowe Schoolcraft (1793–1846). Schoolcraft was also an authority on the North American Indian and was thus a particularly appropriate person to explore the southern shore of Lake Superior. His observations about the various physiographic and geologic features along the southern shore of Lake Superior were the first to arouse serious interest in the region's potential. Fully aware of the vast copper deposits, he argued that federally sponsored surveys would "augment our sources of profitable industry" and "promote our commercial independence."[10] It was not until 1841, however, that things began to move. In that year, the report of the Michigan state geologist, Douglass Houghton (1809–45), confirmed the presence of the region's Precambrian iron and copper and attracted the attention of geologists worldwide. It pointed specifically to the rich copper deposits in the western Upper Peninsula of Michigan and initiated the first copper rush into the area. Houghton observed that the Lake Superior sandstones were comprised of a "Lower Red Sandstone" and an "Upper Gray Sandstone," thus first applying the name red sandstone to the chief sedimentary rock that is visible on the south shore of Lake Superior from Grand Island west to the head of the lake. He noted the apparent location and depth of the formation and the composition of the rocks. He discussed the economic value of the rocks for building materials and concluded that no portion of the Lake Superior region lacked a supply, whether syenites and syenite granites, greenstones, or red sandstone. In regard to the Lake Superior red sandstone, he wrote:

Douglass Houghton (1809–45), 1870s, Alvah Braddish. Oil on canvas. The portrait of Houghton dressed in field geologist clothing places him in front of the Pictured Rocks with his spaniel, Meme. (Courtesy State Archives of Michigan)

> A very good building stone may be obtained from many portions of the lower, or red sandstone formation, and though the cement of this rock is usually not very perfect, yet, frequently, such changes have taken place in the rock, that it has almost taken on the character of granular quartz rock, in which cases, its durability is very much increased. The strata of this rock are usually of a convenient thickness to admit of being easily quarried, and they are so regular that the stone will require but little dressing.[11]

Bela Hubbard (1814–96), assistant state geologist who accompanied Houghton in the expedition to survey the geology of the Upper Peninsula in 1840, predicted the value and the suitability of the sandstones of the Lake Superior region as a building material. He stated: "There is in its building stones a wealth that is hardly yet begun to be realized. No more beautiful and serviceable material than the easily worked and variously-tinted sandstones is found in the West; and her granites, already broken by natural forces into convenient blocks, and as yet untried, will command a market in the time coming, when the solid and durable shall be regarded as chief requisites to good architecture."[12]

Nearly a decade later, geologists for the federal government, David D. Owen and assistants J. G. Norwood, A. Randall, and Charles Whittlesey, examined the public lands along the Wisconsin coast of Lake Superior. In describing the red sandstone of the Apostle Islands, Owen stated: "The ledges are composed of red sandstone with marly deposits. Of the former there is about fifteen to twenty feet exposed above the water level. The lower bed is some seven to nine feet thick and regularly jointed; it would make a fine building material. The color is pleasant—not too red; it reminded me of the rock obtained at the quarry on Bull Run, in the new red sandstone formation on the Potomac, in Maryland."[13]

John Wells Foster (1815–73) and Josiah D. Whitney (1819–96) studied the geology of the Lake Superior region in 1850 and expanded on the earlier reports, adding further impetus to its development. Their two-volume report contained not only factual scientific descriptions of the sedimentary rocks of the copper and iron regions, which they equated to the Potsdam group of New York, but also predictions of the economic value of the stones as a building material and romantic descriptions of their beauty and likeness to the ruins of antiquity. Foster and Whitney concluded that only the sandstones near the trappean belts would furnish good building materials because the stones were cemented with ferruginous matter and appeared to contain fewer impurities that would cause crumbling or discoloration from weathering. They noted that the two- to three-foot-thick bedding and transverse jointing of some of these sandstones would facilitate greatly their extraction into convenient-sized blocks for building. In particular, they observed the bold cliffs that line the shore for several miles above the Portage Entry on the Keweenaw Peninsula that would become the region's major quarry center:

> About a mile above the Portage, a good opportunity is afforded for examination. Here commence a series of bold cliffs, which line the coast for several miles. They are composed of strata of unequal thickness and induration. Some of the strata consist of silex, with thin plates of mica interspersed, while others contain portions of alumine, colored red by

REPORTS

OF

WM. A. BURT AND BELA HUBBARD, ESQS.

ON THE

GEOGRAPHY, TOPOGRAPHY AND GEOLOGY

OF THE

U. S. SURVEYS OF THE MINERAL REGION

OF THE

SOUTH SHORE OF LAKE SUPERIOR, FOR 1845;

ACCOMPANIED BY A LIST OF WORKING AND ORGANIZED MINING COMPANIES; A LIST OF MINERAL LOCATIONS; BY WHOM MADE,

AND A

CORRECT MAP OF THE MINERAL REGION,

DELINEATING THE TOWNSHIP AND SECTION LINES, AND THEIR CONNECTION WITH THE LOCATION LINES;

AND ALSO, A

CHART OF LAKE SUPERIOR,

REDUCED FROM THE BRITISH ADMIRALTY SURVEY.

BY J. HOUGHTON, JR. AND T. W. BRISTOL.

DETROIT:

PRINTED BY CHARLES WILLCOX.

1846.

Frontispiece. From J. Houghton, Jr. and T. W. Bristol, Reports of Wm. A. Burt and Bela Hubbard, Esqs. On the Geography, Topography and Geology of the U. S. Surveys of the Mineral Region of the South Shore of Lake Superior, for 1845 *(Detroit: Charles Willcox, 1846).*

> oxide of iron. The silicious strata afford excellent building materials, and the supply is inexhaustible. Slabs varying from two inches to two feet in thickness, and exposing perfectly level surfaces of forty or fifty superficial feet, can readily be procured. The rock is sufficiently indurated to give it strength, and is little affected by atmospheric agents. The water is of sufficient depth to permit vessels to approach within a few rods of the shore.[14]

This scientific activity was not confined to the Lake Superior region. It was but part of a larger exploration of the country as a whole, an exploration made possible by the dramatic advances in the earth sciences that were a mark of nineteenth-century America. At that time, the emerging disciplines of geology and biology represented the cutting edge of knowledge. They held the same fascination for the nation at large as the exploration of space holds for us today, and their revelations were observed with an acute interest that would lead to a dramatic change in the way Americans saw and understood the land in which they lived.

It was not only physical data, however, that the scientists brought back with them. Their reports also contained vivid descriptions, written with fascination, awe, and reverence toward a varied and spectacular land. More than that, the veracity of these accounts was confirmed by actual visual images provided by artists and photographers who had been taken on the expeditions for no other reason than to record what they saw. The reports of Foster and Whitney on Michigan, for example, fully captured this new attitude toward the unbroken wilderness. In part factual, in part intensely romantic, they contain detailed descriptions of the brilliantly colored rocks that had been shaped by the ceaseless action of the surging lake into striking and beautiful caverns, cornices, grotesque openings and Gothic doorways, arches, and ruinlike shapes reminiscent of antiquity. Equally important, the text is illustrated by lithographs of geological formations that not only have the appearance of architectural forms but are identified as such—Monument Arch, Miner's Castle, the Amphitheater, the Grand Portal, the Chapel. To Hubbard, also, the shapes and forms of the Pictured Rocks near Munising resembled the elements of ancient architecture—buttresses, columns, arches, a dilapidated tower of a time-worn Gothic castle.[15] Charles Lanman (1819–95), writer, explorer, and artist, toured the Great Lakes basin in the 1840s and wrote in the 1850s about the wild and rugged picturesqueness of the sandstone cliffs and clusters at the Pictured Rocks and west of the Apostle Islands, and their resemblance to arches, doorways, and caverns.[16] Forty years after Lanman's romantic accounts, Sam S. Fifield of Ashland described the architectural features of the sandstone cliffs of the Apostle Islands—caverns, grand arches and columns, cathedrals, vaulted chambers, pillars, circular and

Grand Portal, Pictured Rocks, etching by William Hart. From William Cullen Bryant, Picturesque America, *vol. 1 (New York: D. Appleton and Company, c. 1872–74).*

Gothic windows.[17] Thus, by associating the earth's structure with architectural structure, the imagery of geology was made to impress the imagery of architecture, giving wholly new meaning to stone as a building material.

Extracting, Finishing, and Transporting

Companies quarried Lake Superior sandstone as dimension or mill stone, ton stone, and rubble stone. Of these the most marketable and the most costly was dimension stone. Dimension stone was removed in blocks that typically measured eight feet by four feet by two feet in size. The blocks

Arch Rock, Presque Isle, Marquette, photograph 1899. (Courtesy Library of Congress, Detroit Publishing Company Photographic Collection)

were sawn or scabbled and sold by the cubic foot to builders and contractors who cut them into the shapes and sizes specified by their intended position in a particular building. Some companies operated sawmills that cut the stone into square and rectangular blocks known as ashlar and into such architectural components as windowsills. Thus, in 1873 the Erickson Manufacturing Company installed at Marquette a gang of stone saws for the purpose of sawing into different shapes the brown sandstone taken from the quarries in Marquette. Previously all stone from the Lake Superior quarries had been shipped in the rough.[18] Ton stone was broken out in rough, angular, irregular blocks and sold by the ton in pieces weighing from one-half to three tons. Rubble stone was the byproduct of removing dimension stone. It was sold cheaply by the cord in pieces manageable by one or two men and used in building cribs, breakwaters, piers, and foundations. Sometimes rubble was discarded altogether.

Dimension Stone. Portage Red Stone quarry. Portage Red Stone Company, photograph c. 1895. (Michigan Technological University Archives, Adolph F. Isler)

Stone was classified according to its color and suitability for building. Stone uniform in color and free of white blemishes was graded number one, stone not sufficiently free from white to grade number one was graded number two, and stone whiter than the others was graded number three.

Opening a quarry required the removal of the overburden of glacial drift and shale rock, which could cover the sandstone deposit to a depth of up to fifty feet. In this operation, unskilled workers broke up solid portions of the overburden by blasting it with gunpowder placed in drill holes and by wedging it off. They then carted the rubble away. If the sandstone deposit did not stand out in cliffs, workers excavated by blasting or by cutting long narrow channels in the rock in order to free a narrow space for the quarry face. The cost of removing this overburden was a decisive factor in the practicality of opening a quarry at a particular site. It varied with the extent and nature of the overburden, its proximity to the water, and the facilities available for its disposal. Quarry sites at water's edge had a distinct advantage. Here rubble was tossed aside into cribs for docks or shipped elsewhere for breakwaters and foundations. The work of removing the overburden often took place in the winter, but the quarries operated seasonally from April to November. This allowed stone to be removed without excess moisture or frost and thereby ensured good weathering.

Dock. Portage Entry quarry. Portage Entry Quarries Company, photograph c. 1895. (Courtesy Michigan Technological University Archives, Adolph F. Isler)

A quarry was opened after workers stripped the overburden. On the face that the quarry was to be opened, they drilled or, after the 1880s, cut with a steam-powered channeling machine, two parallel channels, about four feet apart, and parallel to the depth of the bed. Workers then broke out a portion of the stone between the two channels and removed a key so as to create a space in which to move other stones as they were broken apart. Quarrymen blasted this break or wedged it out. They either drilled deep holes, larger at the bottom than the top, and charged them with heavy blasts of powder to break the stone apart and move it slightly on its bed; or they drilled two or three holes at right angles to the channels and to the depth of the bed and then cracked or loosened the stone with plugs and feathers. With the latter method, two thin, half-round pieces of soft iron that tapered to a point at one end and were called feathers were placed in a series of holes along the line where quarriers wanted to break a line. The split was made by driving a small wedge-shaped piece of steel called a plug into each pair of feathers and moving along the line gently striking each wedge in its turn until a strain caused the stone to fracture. A derrick lifted out the key.

With the key removed, quarriers could begin extracting the building stone. This was accomplished by a method of channeling and wedging. They removed stone by cutting channels transversely to the face, drilling

Illustration of the method of quarrying sandstone by channeling and wedging. From Plate XXIII of Report on the Building Stones of the United States and Statistics of the Quarry Industry for 1880, *U.S. Department of the Interior, Census Office,* Tenth Census, *vol. 10 (Washington, D.C.: GPO, 1884).*

Quarriers and tramcar loaded with stone blocks. Portage Entry quarries, Portage Entry Quarries Company, photograph c. 1895. (Courtesy Michigan Technological University Archives, Adolph F. Isler)

a series of holes parallel to the face, and then cracking it out with wedges in the same manner by which they had removed the key. The large blocks freed from the quarry were broken into sized blocks by placing a series of iron wedges in a lengthwise groove where a break was desired and striking them repeatedly with sledgehammers. Derricks hoisted the stone directly onto ships or onto tramcars that carried the stone to the dock. At dockside each block was numbered and measured. This information was inscribed on the block and noted in a memorandum.[19]

In 1883 noted Lake Superior quarrier, John H. Jacobs, invented a blasting system that employed a power drill capable of boring any shape hole. Manufactured by the Rand Drill Company in Terrytown, New York, the "Jacobs BB drill" reduced from thirty or forty to two or three the work force needed to blast out stone. Jacobs described how the system, in his opinion "the most perfect quarrying system ever known," worked:

Quarriers and channeling machine. Portage Entry quarries, Portage Entry Quarries Company, photograph c. 1895. (Courtesy Michigan Technological University Archives, Adolph F. Isler)

> We take a block of stone sixty feet or more long, twenty feet wide and ten feet thick and bore a row of holes eight feet apart cutting the corners on the line drawn to break off the twenty foot piece and using about half a pound of powder in the bottom of each hole and putting in the exploders with electric connections and using a blasting battery to set them all off at once, this body of stone could be moved any distance from three or four inches to a foot or more, as you please, by the slightest little amount of powder and by corking the hole two feet from the top by tamping solid to hold the gas and powder until the full strain on that filled the vacuum of the hole. Then it cut the rock in an air line and made a perfect break as directed and the process was a great saving in quarrying the rock.[20]

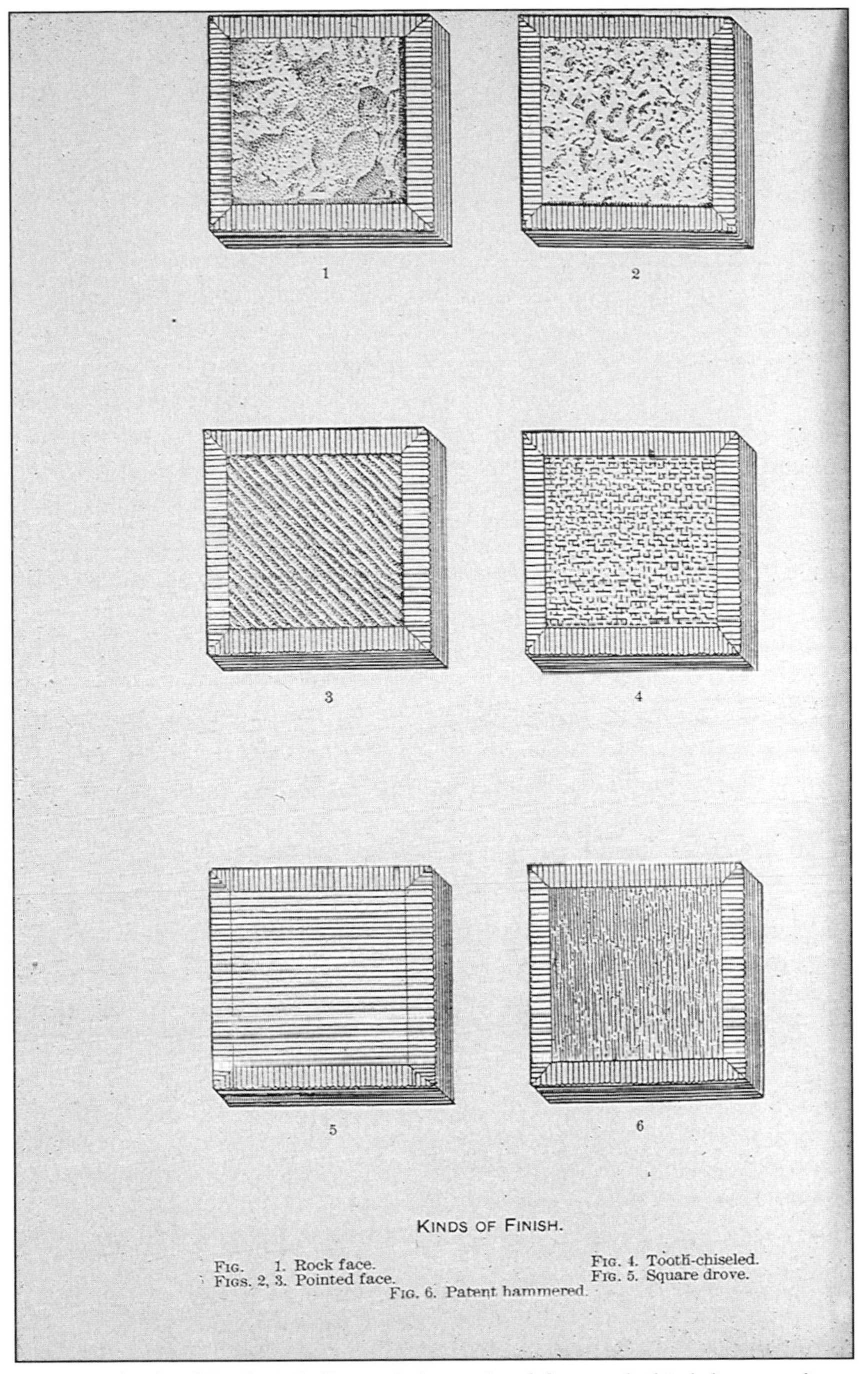

Common kinds of finish, including rock face, pointed face, tooth chiseled, square drove, and patent hammered. From Merrill's The Collection of Building and Ornamental Stones in the U. S. National Museum: A Hand-Book and Catalog, *Annual Report of the Board of Regents of the Smithsonian Institution, Pt. 2 (Washington, D.C., GPO, 1886).*

Quarry companies employed steam drills and steam hoists to extract stone. In the 1890s, for example, the Ashland Brown Stone Company and the Superior Brownstone Company acquired Wardwell channeling machines, which had been invented in 1863 by George J. Wardwell of Rutland, Vermont. This was a locomotive machine driven by steam power that moved over a steel rail track placed on the quarry bed. It carried either a single gang drill that was raised and lowered by levers on one side or two such drills, one on each side. The lifted drill dropped with great force and rapidly creased a channel into the rock. The machine ran backward and forward over the rock while its cutters delivered their strokes, cutting channels in the rock. The machinery employed at many Lake Superior quarries in the 1890s included a gadding machine. The gadding machine undercut and released the block after the channeler had done its work. This machine stood on a platform on trucks that ran on a track. To the boiler, which formed the main support of the machine, was secured a perpendicular guide bar and to this was attached a drill. By raising, lowering, and swiveling the drill, its operator could bore a series of perpendicular holes in any direction. Removal by channelers and gadders, therefore, replaced removal by plugs and feathers as a method for quarrying.

Skillful quarriers took advantage of the systems of natural planes, seams, or cracks of the rocks. Sandstone joints usually run nearly perpendicular to the planes of bedding and descend vertically at similar distances. They tend to intersect in two directions. The first, called dip joints or end joints, runs with the direction of the strata and is inclined from the horizon or the dip or inclination of the rock. The second, called strike joints or back joints, runs transversely at a right angle and conforms in direction to the strike of the rift. Aided by the natural joints of the rock, quarrymen could wedge large blocks off easily.

The stone was then finished to the desired textured finish with hand implements and with saws. Most commonly, stone finishers gave sandstone a rock-face finish, which is the natural face of the rock as broken from the quarry or slightly trimmed down by a pitching tool, or they gave it a sawed-face finish. Often sandstone was dressed with hand implements that textured the surface into finishes known as pointed face, ax-hammered face, patent hammered, brush hammered, square drove, and tooth chiseled. With lathes and chisels, they turned posts and columns.

Suitability as a Building Material

Lake Superior sandstone possessed the strength, durability, and beauty necessary to meet the construction requirements of substantial architecture. It resisted the effects of fire better than most stones and the effects of freezing and thawing as well as most sandstones; it also withstood the extreme temperature changes of the northern climate, retaining solar heat in the winter and insulating against it in the summer. Rich in color and homogeneous in structure, the stone was easily worked, easily adapted to required dimensions, and easily dressed and carved in any fashion. Builders used Lake Superior red sandstone in all kinds of rubble and ashlar masonry and in all parts of buildings, but found it particularly suited to coursed ashlar walling and rock-faced work. The stone was obtainable in large quantities at low cost and was transported cheaply by water. Its primary liability was the presence of gray spots and the seamy nature of inferior stone when not carefully selected.

In the early 1870s, as the quarries were opened, local newspapers extolled enthusiastically the beauty of the reddish brown sandstone, as well as its carvability, durability, and fireproof quality. These accounts also told of the ease by which the stone could be extracted, pointed to buildings erected of the material in the region and in the Midwest, and forecast the economic importance of the sandstone industry to the region. The *Marquette Mining Journal* for 21 September 1872 reported on the sandstone from the quarries in South Marquette that the Marquette Brownstone Company had just acquired from the Wolf and Son Company. The quarry was opened in 1869. The journal stated:

> The stone which has of late been taken out of the Wolf quarry, is as uniform in quality and as easily manipulated in a mass of any size—to an especial pattern if desirable—as if moulded [*sic*] out of soap, aside from the trifle of additional labor required to work the harder substance. This may seem like an exaggeration, but operations on the ground, at the quarry, show that by the usual method of trenching and underseaming any size of stone may be taken out in as perfect a form as it is possible to obtain it without the use of the saw and chisel. Then it is easily worked, so easy in fact that the most experienced stone-cutters decide that from twenty to fifty per cent can be saved in carving it to occupy any place which may be desired in a building. And last of all, it is found that as it is exposed to the atmosphere and the moisture is evaporated it becomes harder than any other stone in use, and when held up in a wall in the same horizontal position it was taken from the quarry, it will not scale, crack or expand, as is the case with much stone used for building.

Three years later, on 20 November 1875, the same newspaper commented on the brownstone from the quarries on Basswood Island in the Bayfield group and in Marquette in the Jacobsville formation:

> Its fire-proof qualities were thoroughly tested in the great Chicago fire, where [L]ake Superior brownstone walls, among them those of the

> *Tribune* building, stood intact, without a crack, scale or blemish being caused by the great heat under the influence of which marble fronts crumbled and fell to the ground. No more beautiful buildings can be imagined than those in this city, the walls of which have been constructed with material from our own quarries; and when we take into consideration its sterling fire-proof qualities, its beautiful color, and the ease and readiness with which it yields to the manipulation of the artisan and builder, the conclusion that it must steadily grow in favor among those who desire, in the construction of buildings, to combine beauty with durability, is inevitable. We regard the sandstone interests of Lake Superior as destined, ere long, to become one of the greatest sources of wealth to the district not only but to those who invest their surplus means in its development.

Nearly twenty years later, as the sandstone industry reached its peak of productivity, the *Ashland Daily Press*'s Annual Edition of 1893 elaborated on the wonderful attributes of Ashland brownstone as a building material:

> Building stone that combines strength, durability and beauty is eagerly sought for by contractors and builders. Ashland brownstone is famed throughout the United States. Its like is found nowhere in Uncle Sam's domain of a rich, brownish red color, indestructable [*sic*] and unchangeable in all kinds of weather, qualities for building purposes equaled by no other stone. Public buildings in Brooklyn, New York, and Milwaukee and other prominent cities are evidence of the estimate in which it is held, while the Ashland government building, the Knight Block, the Wisconsin Central depot and local buildings are standing monuments of its popularity at home.

This boosterism in local newspapers was like that of the local chambers of commerce, but the development of the sandstone industry and the marketing of the stone product needed factual data on the physical properties of stone—its mineralogical composition, texture, hardness, strength, and structure, and the ability of a particular stone to resist mechanical and chemical changes. This information would demonstrate the suitability of Lake Superior sandstone as a durable building material. People and organizations interested in promoting commerce and business on behalf of the sandstone industry called for comparative scientific information. In response, federal and state governments published handbooks and statements that compiled accurate information. The reports prepared by scientists described the physical properties of rocks and their suitability as a building material, the quarries from which they were obtained, and the means available for transporting them. They were intended for popular use by quarrymen, builders, architects, suppliers of building materials, and architectural clients.

In the 1880s the United States Department of Interior published a comprehensive report and statistics on building stones and the quarry industry nationwide. *Report on the Building Stones of the United States and Statistics of the Quarry Industry for 1880* was prepared for the tenth census. Many eminent geologists and engineers described the microscopic structure and chemical analyses of building stones, methods of quarrying them, quarries and quarry regions, stone construction in cities, and the durability of building stones in New York City and vicinity. Six years later, George P. Merrill compiled a handbook and catalog of building and ornamental stones. His *The Collection of Building and Ornamental Stones in the U. S. National Museum: A Hand-Book and Catalog* was published by the Smithsonian Institution in 1886 and later revised and updated. The book catalogues the approximately twenty-nine hundred mineral specimens collected for the Centennial Exposition in 1876 at Philadelphia and was the standard work on stones for building and decoration at that time. In the third edition of 1903, Merrill noted that, since his first edition, the output from and investment in America's quarries had doubled, technology had advanced, and the taste and prosperity of Americans had improved.

Allan D. Conover served as special agent on building stones and the quarry industry for the tenth census. Conover taught civil engineering at the University of Wisconsin and designed many sandstone structures in Ashland and Washburn in the 1890s. He described the quarries of Michigan and Wisconsin. His reports appeared in both the Department of Interior's census report and the Smithsonian Institution's catalog. Conover noted that the Potsdam sandstone would probably furnish the largest quantity and the best building material found within Michigan. Conover described the most desirable of the Lake Superior sandstone as "a rather coarse grained, homogeneous, siliceous sandstone, rather soft when first quarried, easily hewn, but hardening on exposure." He regarded uniformly reddish brown sandstone as a very handsome material for exterior and ornamental work. Even the mottled white or yellowish white and red brown stone, usually rejected, but used in some buildings at Marquette, in Conover's view, was handsome and picturesque. The reddish brown and variegated stones seemed equally durable and reliable when free of the clay pockets that occurred occasionally. Conover noted that stones in the vicinity of the Laughing Whitefish River were "very hard, compact, reddish or speckled," split readily to required thicknesses, and were suited to heavy masonry but too hard for use in ornamental work. The Potsdam sandstone of northern Wisconsin also furnished a handsome building stone. Conover described the composition as of "siliceous grains, medium to somewhat coarse, held together by a cement usually either ferruginous or argillaceous in its character, and . . . generally stained from yellow to deep brown by the ferruginous matter." Like the Marquette stone, the Bayfield stone was plagued with clay pockets that could pit the finished surface or spoil a nearly finished ornamental carving. Bayfield stones, however, were free of the color variations sometimes troublesome in the Marquette stone. Conover reported both stones obtainable in large masses of any size required and easily shipped by water.[21]

In the 1880s and 1890s the Minnesota and Wisconsin Geological Surveys prepared reports of statewide application containing practical advice. In his preliminary report on building stones of Minnesota, W. H. Winchell urged architects and builders to examine rock in its natural bed to ascertain its resistance to changes resulting from alternating heat and cold, moisture and dryness, and to observe the natural outcrop of the rock to determine its permanence of color. Most importantly, he called for the use of locally quarried rather than imported stone because of its appropriateness to the particular environment. Winchell claimed that after a few years of exposure in the cities, Ohio stone would become dirty, dusty, dingy, and disagreeable, but brownstone "maintains its color and always gives an impression of solidity and durability" and "while under the bright and cloudless skies of Minnesota its dark and more somber aspect is far less objectionable than in darker and damper atmospheres."[22] In his 1898 bulletin on the building and ornamental stones of Wisconsin, Ernest Robertson Buckley spoke of the demands and uses of stone, the selection of stone, and the estimated value of stone. He discussed the occurrence, color, composition, adaptability, and strength and durability of the building and ornamental stones of Wisconsin. Michigan lacked comparable handbooks, but state geologists incorporated remarks on building stones in their geological survey and mining reports.

Builders and architects relied most heavily on the results of the crushing strength test and the transverse strength test in determining the comprehensive strength of a stone. They turned to the modulus of elasticity test in calculating the effect of loading a masonry arch or proportioning abutments and piers of railroad bridges subject to stress. In both cases, sandstone tested lower than granite and limestone.

Scientists considered sandstone that had a crushing strength of over 5,000 pounds per square inch strong enough for any modern structure. Buckley showed the results of crushing strength tests on two-inch cubes of sandstone from seven quarry sites in the Bayfield group. The average strength of twenty samples taken was 4,816 pounds per square inch. In Buckley's opinion, the crushing strength of Bayfield sandstone equaled or exceeded that of Jacobsville sandstone and probably equaled that of

Connecticut stone.[23] By comparison, the Michigan commissioner of mineral statistics presented crushing strengths of Jacobsville sandstones: Portage Entry redstone, 7,300 pounds per cubic inch; Marquette raindrop variegated, 10,780; and L'Anse brownstone, 10,645.[24]

Buckley stated that testing to determine the resistance of stones to the effects of freezing and thawing were inconclusive. Although tests placed the porosity and ratio of absorption of Bayfield sandstone at 4.5 to 11 percent its weight, a ratio that could exceed the 10 percent considered vulnerable to injury from the conditions of alternating freezing and thawing, other factors affected the ability to resist weather. Also a factor is the rapidity with which stone gives off water. This depends on the size of the pores. Buckley concluded that the pores of Lake Superior sandstone were sufficiently large to release quickly rather than retain the water that might collect in them.

In some cases, changes produced by weathering improved the stone. Sandstones undergo a process of hardening on being removed from the quarry. The water that permeates the stones holds in solution or suspension a small amount of siliceous, calcareous, ferruginous, or clay-like matter. On exposure to the atmosphere, capillary action draws this "quarry water" to the surface of the block where it evaporates. The dissolved or suspended material is deposited at the surface and serves as an additional cementing constituent to bind the grains more closely together. Thus, carving was best done immediately after extracting the stone while the "quarry water" was still present to form a crust rather than after the structure was completed.

Buckley claimed that tests demonstrated that Lake Superior sandstone resisted heat effectively. Up to a temperature of eight hundred degrees Fahrenheit, it was little injured. Not until subjected to twelve hundred degrees Fahrenheit, a temperature no stone can endure, did it crack, crumble, and lose its color and strength. By comparison, granites and limestones change more rapidly at lower temperatures.

One observer who inspected the ruins of Chicago after the fire of 1871 reported in the *Marquette Mining Journal* that only Marquette red sandstone and Akron stone withstood the melting heat of the fire. All others cracked and crumbled before the scorching flames. "The stone of Marquette . . . remains intact in places amid the hottest heat and where it was tumbled down by the warping of bricks or the giving away of iron it can be seen unscorched and merely fractured by the weight of its descent."

The reporter went on to say that the fireproof quality of Marquette sandstone was one of its most important attributes as a building material: "No better argument in favor of the great value of Marquette stone for building purposes than this could be given and we predict for it a very extensive sale in the future."[25]

Sacred Heart Church, 1894, Demetrius Frederick Charlton, L'Anse, photograph c. 1895. With its variegated L'Anse sandstone church, this Catholic congregation declared the values of its institution and the social position of its members. (Courtesy Michigan Technological University Archives)

Color and texture determined to a large extent the market value of the stone. Many thought top quality purplish brown Marquette stone the most handsome stone quarried on Lake Superior because of its "raindrop effect," which made it appear as though a few heavy drops of rain had fallen on it and the moisture partly absorbed. The rich brown color of

Vertin Brothers Department Store, 1885, Calumet, photograph c. 1895. Variegated red and white Portage Entry sandstone, rough cut and rubble, finishes the exterior walls of the store established by Austrian Americans John and Joseph Vertin. Two more stories were added in the later remodeling to the plans of Charles W. Maass. (Courtesy Michigan Technological University Archives)

sandstone was well suited to cities and industrial centers since it would become neither dingy looking nor discolored as would white marble, gray granite, and light grayish yellowish brown Ohio sandstone. In particular, it seemed suited to the iron range cities where red iron ore dust covered everything.

In the early 1890s, clear Marquette brownstone sold for $1.30 per cubic foot. At the same time, top grade Portage Entry redstone sold for $1.20 per cubic foot. These prices were three times greater than those of second quality or variegated stone. Variegated stone, sometimes known as bacon stone for its red and white streaks, was less desirable but more common. It sold for $0.40 per cubic foot. At Portage Entry, second quality variegated stone with white bands running through it was taken out only on order and sold for $0.60 per cubic foot.[26] Rubble stone and inferior stone

Andrew A. and Laura Greenough Ripka House, 1875, Carl F. Struck, Marquette, photograph c. 1985. With this Gothic Revival house on Arch Street, the Philadelphia-born-and-educated manager of the railroad company dock and speculator in mines proclaimed the position of his family in Marquette society. (Courtesy Balthazar Korab)

was suited only for foundations. In some cases, builders laid it in the front exterior walls of buildings. Substituting inferior stone for top grade stone could lead to the impression that all Lake Superior sandstone was poor and damage the reputation of the sandstone.[27] Many public buildings, banks, churches, schools, and houses in the Lake Superior region employed first grade sandstone. Frequently, a variety of grades was used in a single structure. For example, the Edward Ryan Block (1898) at Calumet was constructed of different grades of Portage Entry sandstone: the front is number one grade, uniformly red in color with alternating courses of rough and smooth dressed blocks at the first story level, with coursed ashlar at the second level, and with random ashlar at the third level; the side walls are number two grade, a heavily striated red and white stone.

Promoting and Marketing

Owners, developers, and managers of quarries, sometimes assisted by commissioners of mineral statistics, vigorously promoted their stone in a variety of ways. Initially they overcame uncertainties about the stone's soundness with statements from scientists, geologists, and engineers, who analyzed its composition and tested its strength and porosity. Architects, builders, and clients who had used the stone furnished the quarry companies with statements about the stone's beauty and the ease with which it could be carved and worked. Quarry companies distributed samples of their product to cut-stone dealers in Cleveland, Chicago, Detroit, Saint Paul, and other midwestern cities. Eventually quarry companies opened offices and dealerships in the Lake Superior region and actively marketed the stone directly among local architects and builders. The Portage Entry Quarry Company compiled a pamphlet illustrating buildings constructed with Lake Superior stone for distribution as prospectuses or advertisements. Physical evidence of the fashionableness of prominent Bayfield and Jacobsville sandstone architecture in large midwestern cities enhanced the material's reputation.

The Basswood Island Brownstone Company, the first company to open a quarry in the Bayfield area, initially experienced difficulty in winning acceptance of its product in the market. However, with statements from scientists and experts about the durability of the stone for building purposes and with the Milwaukee County Courthouse in place, the company and the industry were better able to promote the soundness of their product.

Timothy T. Hurley of the L'Anse Brownstone Company refuted earlier reports that L'Anse stone was blemished with white specks and promoted stone from his quarry with firsthand accounts of a Chicago architect and a

Chicago builder who had inspected the L'Anse Brownstone Company's quarry in the summer of 1876. O. L. Wheelock and George A. Gindele enthusiastically spoke about the quality and quantity of the L'Anse sandstone and noted that the stone on the dock was unblemished. The *Marquette Mining Journal* for 19 August 1876 summarized Wheelock's and Gindele's reactions as follows: "They pronounced the stone equal to if not superior to any they have seen and will return to Chicago prepared to recommend it to all who wish to combine beauty with durability in the construction of residences or public buildings." Gindele Brothers, contractors in Chicago, stated: "This is to certify that we have used considerable quantities of your Brown Stone for the past two years and it gave satisfaction in almost every instance. We consider it a very durable stone and cheerfully recommend it. The knowledge we have of your quarry we consider that any required size can be furnished."[28] Wheelock, architect, of Chicago said: "Having made a thorough examination of your Brown Stone quarry at L'Anse, I feel confident that it is the most desirable stone that has yet been placed on this market. All things considered it has fewer defects, and is more uniform in color than other brown stone, and can be obtained in any size required. I saw one piece of the following dimensions split from the main rock—10 feet thick, 12 feet wide, 34 feet long and very nearly perfect in every particular. I am satisfied that a large amount of it would be used in this city next season, provided it could be obtained when wanted."[29]

In 1877 Charles E. Wright, Michigan Commissioner of Mineral Statistics, assisted the industry as a whole by seeking the opinion of architects in Detroit, Chicago, and Cleveland on the value of Marquette brownstone and publishing their testimony in his annual report. William Scott and Company, architects and civil engineers in Detroit, testified: "We have used the Marquette Brownstone, and consider it to be of the very best quality for building purposes, fine grained, easily cut, and keeping color, while the stone hardens with age and exposure to the atmosphere. We consider it fully equal in quality to the Portland Brownstone." Cudell and Blumenthal, Chicago architects who had designed the red Lake Superior sandstone Cyrus McCormick House (1877) claimed:

> Having used [Marquette brownstone] to a considerable extent, we consider it an excellent building stone, especially adapted to elegant residences and structures, where its beautiful color appears to an advantage. This latter feature must ever secure it favor with the true artist, being rich and warm, and suggestive of vitality. As a working material, it is all that the artist can desire, being soft and tenacious, thus enabling the stone cutter to cut and carve the most elaborate and intricate designs of the architect. Durability being one of the essential qualities, we would

> add that having carefully observed buildings constructed from this stone which have stood for a number of years, that in our opinion it possesses this property in a high degree. During a short visit to the quarry last summer, we were highly pleased with what we saw, and the inexhaustible quantity of brownstone.[30]

Testimony from these leading architects lent credence to the claim that Marquette brownstone was suitable as a unique and sound building material.

Even when the market for red sandstone declined after the turn of the century, quarry companies expected their quarry superintendents to aggressively market locally the stone that remained in supply. Thus, the secretary of the Portage Entry Quarries Company sent instructions from the main office in Chicago to the quarry office at Jacobsville to promote the stone in Calumet. "We note that you saw Paul Roehm and that he is figuring on a job that will take all the stone that is at Traverse Bay; we do hope that you will sell him the stone and clean it all up." Later the same year, the secretary again wrote to the Jacobsville office: "We also note what you say in regard to the church that is to be built at Calumet and we shall certainly expect you to look after this and sell them the stone we have there [at Marquette]."[31]

J. W. Wyckoff (1845–1934), superintendent of the Portage Entry Quarries Company's quarry at Jacobsville, sought statements from architects, cut-stone contractors, and clients in Duluth, Minneapolis, and Saint Paul on the desirable qualities of stone from the Traverse Bay Redstone Company quarry on the Keweenaw Peninsula.[32] In fact, Wyckoff maintained a list of midwestern architects, contractors, and clients expressing satisfaction with the use of Portage Entry red sandstone. Lowell A. Lamoreaux, a Minneapolis architect, stated: "'The Portage Red Stone' furnished by you on several pieces of work recently completed has been more than satisfactory. I have never seen building stone so free from imperfections of any character as the stone furnished from your quarries has been. I shall continue to use this stone where ever it will require a good quality of red stone." Frank L. Young and Carl E. Nystrom, architects and superintendents in Duluth, wrote: "My experience with your stone is that it is the finest stone known to me, of all the stone around the lakes it is the finest in quality and the most uniform in color." John A. Young and Sons, cut-stone contractors in Saint Paul, claimed: "During the past season we have used about 4000 cu ft of your 'Portage Red Stone.' We are entirely satisfied with it. Our customers, also specified same; are well pleased. We have heard no word of complaint from anyone about either quality or color, but have heard architects and contractors speak favorably of it." In addition, W. P. Arcking, who used the stone in his house at 1800 Vine Place, Minneapolis;

Jones and Hartley, stone contractors in Minneapolis; and H. K. Jennings, cashier of the Merchants' National Bank in Charlotte, Michigan, wrote letters expressing their satisfaction with the stone.[33] Architects, contractors, and scientists generously praised Lake Superior sandstone and made known their positions publicly, giving prospective buyers firsthand statements from recognized public figures.

The Portage Entry Quarries Company promoted its "Portage Red" sandstone and "Marquette Raindrop" sandstone in a forty-eight page pamphlet published about 1900. This booklet contained an analysis of the stone's suitability as a building material prepared by F. H. Denton, instructor, Michigan Mining School. Denton reported on the results of mechanical tests of compression conducted at the Watertown Arsenal in Massachusetts and concluded:

> As a result of this investigation it would appear that the Portage Entry Sandstone is well adapted to building purposes. Its chemical composition is such as to indicate no danger of the stone losing its color or disintegrating from any chemical changes in its composition. Its crushing strength is above the average for sandstones, and yet not sufficiently high to cause difficulty in working. Its unusual uniformity of texture increases its value both as regards durability and appearance. The only agent likely to affect it appreciably is the alternate thawing and freezing of some climates. This agent is the most powerful one and no stone can withstand its action indefinitely. The Portage stone has about the same ratio of absorption as many other sandstones that have withstood the action of frost well. Since the ratio of absorption is indicative to a certain degree of the power of stones to resist the weather, it is reasonable to expect the Portage stone to weather as well as other sandstones of similar composition. If the stone after quarrying should be well seasoned before being exposed to the weather in buildings, its durability would probably be still further increased.

The booklet contained full-page illustrations of buildings constructed of stone from the Portage Entry Quarries Company and listed a hundred more buildings located in sixteen states—eleven were midwestern states—and in two provinces of Canada. It cited fifteen examples in the Lake Superior region: ten structures in Duluth, three in West Superior, the Gogebic County Courthouse at Bessemer, and the Michigan Mining School at Houghton. It contained letters from one geologist, three architects, and the president of the Northwestern Guaranty Loan Company testifying in support of the stone's desirability as a building material. The booklet indicated that the main office of the Portage Entry Quarries Company was in Chicago and that branch offices in Cleveland, New York, Philadelphia,

Pittsburgh, Duluth, Saint Louis, and Toronto distributed stocked supplies of the red sandstone.[34]

Quarries

The major red sandstone quarries were clustered in three geographic locations and were scattered along the south shore of Lake Superior where the stone crops out. Quarries were centered at Marquette and at Keweenaw Bay in the Jacobsville formation and at Chequamegon Bay in the Chequamegon formation of the Bayfield group. Quarries were scattered along the lakeshore where the Rock, Laughing Whitefish, and Salmon Trout Rivers empty into Lake Superior in Alger and Marquette Counties, Michigan, in the Jacobsville formation, and where the Cranberry, Flag, Iron, and Amnicon Rivers flow into the lake in Bayfield and Douglas Counties, Wisconsin, in the Orienta formation of the Bayfield group. On the Saint Louis River south of Duluth at Fond du Lac were three quarries.

Although mentioning many quarries and companies, this section discusses in some detail three operations, one from each of the primary clusters of quarry activity—Marquette, Keweenaw Bay, and Chequamegon Bay. See appendices 1 and 2 for information on significant quarry companies, their quarry locations, and stockholders.

Marquette

A portion of the Jacobsville formation extends in a three- to four-mile-wide belt along the south shore of Lake Superior from L'Anse to Munising. As early as the 1860s, builders of the Schoolcraft and Bay iron blast furnaces in Alger County opened quarries at the southwest end of Grand Island and at Powell's Point on Grand Island Bay. Later, at the points where the Laughing Whitefish and Rock Rivers empty into Lake Superior and break through the cuesta, three companies extracted sandstone. Craig and Wagner briefly quarried a dark brown and variegated sandstone at Laughing Whitefish River in the 1870s, and the Rock River Brownstone Company and the Butler Brownstone Company quarried at Rock River in the 1890s.

Between 1869 and 1900, more than fifteen companies intermittently extracted sandstone from quarries in Marquette County. The principal quarries were located at two sites south of the center of Marquette where patches of Jacobsville sandstone are exposed: at South Marquette and at Mount Mesnard. Two other quarries were opened where the Jacobsville formation crops out along Lake Superior: six miles northwest of Marquette at the mouth of the Salmon Trout River, and thirteen miles northwest of

Marquette and three-quarters of a mile north of the Little Garlick River at Thoney's Point.

John H. Jacobs and the Wolf and Sons Company; Furst, Jacobs and Company; Marquette Brownstone Company; and Wolf, Jacobs and Company

At South Marquette near Lake Superior in Section Twenty-six, Township Forty-eight North, Range Twenty-five West, companies formed and operated by Chicago, Detroit, and Marquette investors, several of whom were cut-stone contractors and dealers, and themselves trained as stone workers in Germany and Ohio, extracted stone. Eventually most of the companies restructured and consolidated into one firm, the Marquette Brownstone Company. The Marquette Brownstone Company became the major quarry company at Marquette. John H. Jacobs supervised this firm and its predecessors and invested in, developed, and managed several other companies, all top producers. Indeed, Jacobs eventually would play a principal role in the development of the sandstone industry in the Lake Superior region.

In the 1860s a quarry had been partially opened in Section Twenty-six on the J. P. Pendill Farm. From it, stone had been taken for the construction of Marquette's earliest substantial business blocks—in fact, the only buildings that survived the fire of 1868—and for the construction of the Marquette and Ontonagon Railroad shops.

In 1869 Peter Wolf (1819–1902), a German-born Chicago cut-stone contractor, acquired the opened quarry in a portion of the Pendill farm from George Craig.[35] Recognizing the economic value of the sandstone in South Marquette, Craig had speculated in the Pendill farm site in April 1869 and had even taken out some rough stone from the site for several contracts.

The *Marquette Mining Journal* for 28 May 1870 also thought the large bed of excellent red sandstone on this land was valuable. Proper development would demonstrate its immense value. It seemed that the quality of this stone surpassed that used extensively in fashionable quarters of New York City and surpassed any known deposits in the country. Fine grained, compact, clear brown, and free from streaks or blemishes, the bed seemed inexhaustible.

While conducting a geological survey of the eastern portion of the Upper Peninsula in 1871–73, Carl Rominger investigated the sandstones of Marquette. He described the quarry site in a sandstone patch at Marquette as follows:

> The Potsdam deposits seem to have formed a continuous belt all around the Huron mountain district, which must have been an island in the ancient ocean. But a part of these deposits has been washed away again,

> and only in protected situations have patches of the rocks resisted destruction in places where the denuding forces had freely acted. One of these patches, surrounded by Diorite and Slate hills, we find within the city limits of Marquette, in a small side valley, at the lake front of which the Marquette gas works and an iron furnace have been erected.[36]

From the quarries opened in this recess, where the stratification was regular, quarrymen obtained a fine building material.

The Peter Wolf and Son Company was the first to extract stone from the Marquette quarry by using plugs and feathers to trench and underseam any size stone in a form nearly as perfect as that obtained with the saw and chisel. The earlier method of extracting stone by blasting with powder frequently checked, shattered, damaged, and wasted the stone. At first Wolf and Son shipped most of its fine large blocks of uniformly dark brown, moderately coarse-grained sandstone to Chicago. It sold the rubble cheaply for local use in cellar walls and foundations and also supplied the local market with pieces of coarser-grained, less uniformly colored sandstone for home consumption. People anticipated that the development of the South Marquette quarry by Wolf and Son would greatly enhance the growth of business and commerce in Marquette. Marquette people expected to soon supply all of the Great Lakes cities with the best building material in addition to furnishing the nation with iron.[37]

Peter Wolf employed Jacobs as foreman of his quarry in April 1870. In fact, Jacobs came to Marquette from Ohio to take the job. Since age eleven he had worked in the stone quarries at Independence and Amherst, Ohio. Jacobs was born in Lorain County, Ohio, the son of German immigrants from Saxony Weimer. Once in Marquette, his supervision of the Wolf quarry was interrupted briefly each winter for two years beginning in October 1870, when he returned to Ohio to earn a business degree at Oberlin College. In December 1872 he married Peter Wolf's daughter, Mary. Soon after returning to Marquette to resume his duties at the quarry fulltime, Jacobs began acquiring a financial interest in the business and would acquire an interest in its successor, Furst, Jacobs and Company. Later, in the 1890s, Jacobs would establish and run the Kerber-Jacobs Redstone Company to extract stone at Portage Entry. Jacobs also served as mayor of Marquette for three terms.[38]

But in September 1872, Wolf and Company was sold to the recently incorporated Marquette Brownstone Company. This company had purchased the remaining Pendill farm parcel adjacent to the Wolf quarry from George Craig three months earlier and opened its quarry.[39] In the sale the Marquette Brownstone Company purchased from Wolf and Company for twenty-six thousand dollars "all the horses, wagons, tools, derricks and personal property of every description connected with the stone quarry at

John H. Jacobs (1847–1934), Bailey and Tooker, photographers, Marquette, 1921. (Courtesy Marquette County Historical Society)

Marquette, and oats on the way and all the stone quarried and on dock at Marquette belonging to us . . . everything owned by the firm in Marquette (except the stone sold to Wyckoff and to Green and two pieces of stone belonging to J. H. Jacobs)."[40]

Jacobs, however, continued to supervise the operations of the Marquette Brownstone Company as it worked the former Wolf quarry and expanded to the adjacent site in Section Twenty-six. The principal stockholders of the Marquette Brownstone Company were Peter White,

George Craig (1819–92), photograph date unknown. (Courtesy John R. Stuart Estate, Jane Ball and Charles K. Stuart)

Samuel P. Ely, William Burt, Frederick P. Wetmore, Sidney Adams, and Henry R. Mather, all frequent collaborators in local and regional business ventures. Jacobs had a moderate interest, and Alfred Green, a local builder, had a small interest. Two experts in Upper Peninsula geology, Thomas B. Brooks of the economic division of the state geological survey of the Upper Peninsula, and Raphael Pumpelly, who taught mining engineering at Harvard and who had served as state geologist of Michigan from 1869 to 1871, held small interests, probably given to them by primary investors to lend scientific credibility to the venture and, thereby, to encourage the investment of others.[41]

The Marquette Brownstone Company developed the quarry with new energy. In 1874 the company built a stone sawmill, and in 1875 it constructed a large pier, sturdy enough to meet the special needs of handling heavy blocks of stone with derricks and loading them onto ships. The rock-filled approach ran on a moderate incline from the shore to the bridge and pier that stood in sixteen feet of water. The quarry work force averaged about forty employees but reached sixty-three in 1878.

Production at the Marquette Brownstone Company quarry rose steadily from the extraction of 26,390 cubic feet of sandstone in 1869 to 69,000 cubic feet in 1877. The company promoted the use of sandstone locally. In a flurry of activity, the company furnished stone locally for the Superior Building (1871–73), the High School (1874–75), and Saint Paul's Episcopal Church (1874–76). Certainly Peter White, a vestryman at Saint Paul's Church, and the other major stockholders, all prominent Marquette citizens, influenced the choice of sandstone for the church and school. White and Mather selected the sandstone for their own building, the Superior Building, built by Alfred Green who held a minor interest in the company. The Marquette Brownstone Company sold the stone in the Lake Superior region at Houghton, Champion, Escanaba, and Sault Sainte Marie. To reach a broader regional market, S. Brownell, manager of the quarry, traveled out of town to establish agencies for the sale of the stone at Milwaukee, Detroit, Chicago, and elsewhere. In 1876 and 1877 building stone was shipped to these cities as well as to Erie, Pennsylvania; Sarnia and Prince Arthur's Landing, Ontario; Oshkosh, Appleton, and Racine, Wisconsin; Louisville, Kentucky; and Cleveland and Dayton, Ohio.[42]

In 1878 the Marquette Brownstone Company leased its quarry operation to Edward M. Watson and E. B. Palmer, a local firm that supplied goods, supplies, and materials to lumbermen, furnace men, and townspeople. Jacobs continued to oversee the operation for Watson and Palmer until 1881.

Meanwhile, in October 1872, the Burt Freestone Company opened a quarry on property adjoining the Marquette Brownstone Company's quarries. Although the deposit on the Burt site resembled that on the Marquette Brownstone Company site, the stone was lighter in color and inferior in quality. In time, the Burt Freestone Company erected a sawmill and installed machinery to cut the stone into window sills, water tables, columns, and caps. It sold these pieces along with blocks and ashlar in brown, variegated, and mottled colors. The company was less successful than others and ceased operations by 1879. In 1883 Wolf, Jacobs and Company leased the Burt quarry from the Burt family and paid the Burts royalties for the stone the company extracted.

BEARD'S DIRECTORY OF MARQUETTE COUNTY. 27

MARQUETTE
Brownstone Company

S. P. ELY, President. PETER WHITE, Treas.
M. H. MAYNARD, Sec'y and Gen'l Manager.

Own and Operate the Celebrated Quarries of

RED FREESTONE

Very Soft and Rich in color, easily worked when new, and becomes very hard upon exposnre. Is unchanged by the] elements and remarkably durable in color.

Address, M. H. MAYNARD, Gen'l Manager,
MARQUETTE.

HURON BAY
Slate and Iron Company.

W. L. WETMORE, President. M. H. MAYNARD, Sec'y and Treas.

PRODUCE A CLEAR

Black Slate

Of very fine texture and lustre, equal to any in the world.

QUARRIES ON HURON BAY.

General Office at Marquette, Mich.

Address, M. H. MAYNARD, Secretary.

Advertisement for Marquette Brown Stone Company and Huron Bay Slate and Iron Company. From Beard's Directory and History of Marquette County *(Detroit: Hadjer and Bryce, Steam Book and Job Printers, 1873), 27.*

Burt family members of Detroit and Marquette filed articles of incorporation for the Burt Free Stone Company in 1872. The company had a capital stock of five hundred thousand dollars. The Burt family long had engaged in identifying and promoting the mineral resources of the Upper Peninsula. John and William Burt had assisted their father, William Austin Burt (1792–1858) in a ten-year land survey of the Upper Peninsula, conducted in coordination with Houghton's geological survey. With the knowledge and skill gained in doing the land survey, John Burt (1814–86) furthered the development of the mineral resources of the Upper Peninsula through the promotion of the Lake Superior Iron Company, the canal at Sault Sainte Marie, and several iron manufacturing companies. He served as president of the Burt Freestone Company. William Burt (1825–92) speculated in land, iron mining and manufacturing, invested in Marquette businesses, and managed the Marquette and Pacific Rolling Mills. John Burt's son, Hiram A. Burt (1839–1921), manager of several iron companies and investor in real estate, joined in the family company. Two of William Burt's sons, A. Judson Burt and William A. Burt (1851–93), cashier of the Hurley National Bank at Hurley, Wisconsin, also participated.[43]

Like the Marquette Brownstone Company, the Burt Freestone Company promoted the use of brownstone locally. As a member of the Baptist church, William Burt probably influenced the decision of its building committee to use variegated brownstone for its Gothic Revival structure in 1884 in Marquette. And Hiram Burt chose smooth-cut hexagonal blocks of variegated brownstone for his elegant Second Empire house on Ridge Street in 1872–76.

In 1883 John H. Jacobs persuaded Peter Wolf and Son to form Wolf, Jacobs and Company to operate and to open the Burt Freestone quarry more extensively, even though some thought the quarry to be exhausted. Wolf, Jacobs and Company built a mill to dress for market stone from both its Marquette quarries and a quarry that it had opened by then at Portage Entry on property leased from the Lake Superior Brown Stone Company. By 1887 Wolf sold out his interest in the company, and it became Furst, Jacobs and Company. The Furst family of Chicago provided additional capital needed for machinery, equipment, real estate, and expansion.

Henry Furst and Company had been established as a cut-stone contractor in 1861 and was restructured in 1885. The partners included Henry Furst (1835–?), a German immigrant who had learned the stone cutter's trade in his birthplace, Ottweiler, near Saarbruck, Germany, and had practiced it in Cleveland and Chicago, before forming a partnership with Henry Kerber in Chicago in 1861. Eventually Furst was joined by his nephew, Peter W. Neu (1846–?), also a German-born-and-trained stone cutter, and by his son, Henry Furst, Jr. (1863–?), a graduate of Yale College.[44] Then the company expanded even further, not only working

Furst, Neu and Company quarry at South Marquette. From **Marquette, Mich. and Surroundings, Illustrated** *(Chicago: Pettibone Press, 1896?), 59.*

quarries formerly operated by the Marquette Brownstone Company and Burt Freestone Company, but also, in 1891, opening new quarries on land it acquired in the First Ward.

By 1890 Furst, Jacobs and Company showed substantial profits. It divided $100,000 in cash dividends among six equal partners. It paid $500,000 in expenses, $60,000 in wages, $129,290 in freight, and $31,181 in royalties. Its assets included accounts receivable of $133,000, machinery worth $75,000, Marquette real estate valued at $30,000, and capital investments of $180,000.[45] In 1891 Jacobs sold his interest in Furst, Jacobs and Company so as to concentrate his effort on developing and operating the Kerber-Jacobs Redstone Company at Portage Entry. Furst, Jacobs and Company restructured as Furst, Neu and Company and consolidated with Portage Entry Red Stone Company as Portage Entry Quarries Company. That operation is the subject of the next section.

Keweenaw Bay

The profitable extraction of sandstone in the Keweenaw Bay area occurred where the Jacobsville formation crops out on the shores of the bay. This outcropping extends in a belt from Pequaming southwest down the east shore of Keweenaw Bay to L'Anse, around the head of the bay to Baraga, and northeast up the west shore of the bay to a point on the Keweenaw Peninsula several miles north of the entry to the Portage River. The rock at L'Anse is a hard sandstone of a duller purplish brown than Marquette brownstone. The stone on the east coast of the Keweenaw Peninsula and at Portage Entry is fine- to medium-grained and strong red. Although the stones were said to be equally durable, builders preferred Marquette over L'Anse sandstone because of its brighter color and its carvability. Over twenty companies owned or leased sites and operated quarries intermittently in the two major locations at L'Anse and Portage Entry and at several scattered locations in Baraga, Houghton, and Keweenaw Counties between 1875 and 1918. Production peaked during the late 1880s and early 1890s, when the Portage Entry Quarries Company, which operated its largest quarries on the west shore of the bay, probably became the largest producer of sandstone in the Lake Superior region. Production declined in the early 1900s and ended altogether before World War One.

John H. Jacobs and J. W. Wyckoff and the Wolf and Jacobs Company; Furst, Jacobs and Company; Furst, Neu and Company; Portage Entry Quarries Company; Kerber-Jacobs Redstone Company; and Traverse Bay Red Stone Company

At Portage Entry, fourteen miles southeast of Houghton, companies opened their famous quarries in a twelve-foot bed of sandstone. Recognizing the superb quality of the sandstone and the excellent facility for its shipment by water, speculators acquired titles and leases to land on the northeast side of the entry as early as 1870. At this point, the stone crops out horizontally in a bluff that rises from the bay and is readily accessible for removal to ships. But quarry operations did not begin here until 1883. Around them grew the settlements of Jacobsville, also known as Craig and Portage Entry, and Red Rock. Two of the largest companies consolidated in 1893 and gradually took charge of some of the smaller ones as well. The largest producers of sandstone in this area were the companies associated with John H. Jacobs from 1883 to 1902: the Wolf and Jacobs Company and its successors, the Furst, Jacobs and Company, and the Furst, Neu and Company; the Portage Entry Quarries Company, a consolidation of the Furst, Neu and Company and the Portage Entry Red Stone Company; and the Kerber-Jacobs Redstone Company.

In 1883 Wolf and Jacobs Company opened the first quarry at Portage Entry in Lot One of Section Nineteen, Township Fifty-three North, Range Thirty-two West, on the twelve-foot-deep sheet of sandstone discovered by Jacobs and Craig more than ten years earlier and noted by Foster and Whitney more than thirty years earlier.

George Craig (1819–92), keeper of the lighthouse at Portage Entry from 1876 to 1886, reportedly struggled to organize a company to operate a quarry in the bed of red sandstone just northwest of the lighthouse. Whether or not his efforts eventually led to Wolf and Jacobs Company and

Furst, Jacobs and Company opening the quarry is unclear. What is known is that Craig sailed or coasted in a small boat along the south shore of Lake Superior from Laughing Whitefish Point to Ontonagon searching outcrops of redstone to identify suitable quarry sites. A stone mason by trade, Craig had constructed many early industrial structures in the Marquette area.

The Wolf and Jacobs Company leased Lot One from A. Kidder of Marquette and the Lake Superior Brownstone Company.[46] At this site, the clear red stone was considered unsurpassed for building purposes and inexhaustible. Even the laminated red and white stone was thought suitable for trimmings and ornamentation. The *Marquette Mining Journal* described the sandstone of Lot One as follows: "The thick layer of uniform merchantable stone, of beautiful color and fine texture, rises out of the lake at the western boundary of the property, and runs a quarter of a mile east where it thins out abruptly and becomes worthless. At the western limit overlying the clear stone is ten feet of thinly laminated stone, much of which is very beautiful variegated and striped, clear white alternating with clear brown."[47]

The first season over forty men worked to open the quarry. The next season Wolf and Jacobs Company built a pier dock formed of cribs that was substantial enough to hold the tremendous weight of the stone. The following year the dock was extended to stretch 450 feet into the lake.

With the investment of Chicago men, the company reorganized as Furst, Jacobs and Company in 1887. John H. Jacobs supervised the work at the Lot One quarry during the years of greatest production and expansion, eventually selling out his interest in 1891. The *Marquette Mining Journal* for 6 June 1891 reflected that with no money but with a lot of hard and intelligent work, the firm of Wolf and Jacobs Company began to develop a reputation, and, after being succeeded by Furst, Jacobs and Company, assumed enormous proportions. From its early production of 90,000 cubic feet of stone in 1885, the company effected technological improvements so that production consistently exceeded 300,000 cubic feet annually between 1887 and 1891. Production peaked at 450,000 cubic feet in 1890.

With Jacobs's departure in 1891, the Furst, Jacobs and Company reorganized as Furst, Neu and Company. In 1892 it employed seventy-five workers in extracting 228,643 cubic feet of block stone. Two years later, in 1893, Furst, Neu and Company and the Portage Entry Red Stone Company consolidated as the Portage Entry Quarries Company, with capital stock of one million dollars. The firm placed J. W. Wyckoff, a Marquette building contractor and dealer in building materials, in charge and opened an additional quarry on Section Eighteen, Township Fifty-three North, Range Thirty-two West. Wyckoff continued to manage

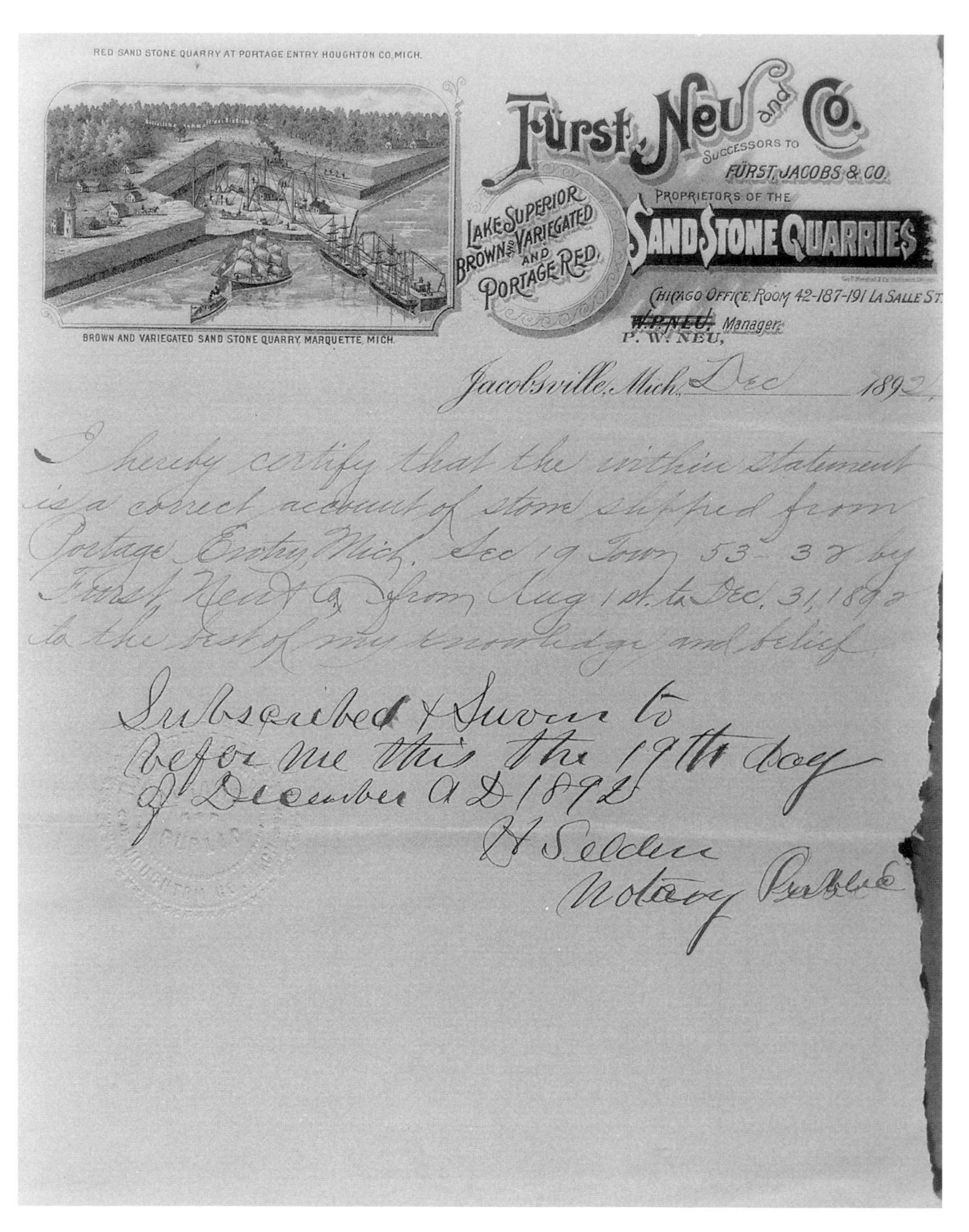

RED SAND STONE QUARRY AT PORTAGE ENTRY HOUGHTON CO. MICH.

BROWN AND VARIEGATED SAND STONE QUARRY, MARQUETTE, MICH.

Fürst, Neu and Co.

Successors to Fürst, Jacobs & Co.

Proprietors of the

Sand Stone Quarries

Lake Superior Brown and Variegated and Portage Red.

Chicago Office, Room 42-187-191 La Salle St.

~~W. P. Neu,~~ Manager.
P. W. Neu,

Jacobsville, Mich. Dec 1892

I hereby certify that the within statement is a correct account of stone shipped from Portage Entry, Mich. Sec 19 Town 53 - 3 W by Fürst, Neu & Co. from Aug 1st. to Dec. 31, 1892 to the best of my knowledge and belief.

Subscribed & Sworn to before me this the 19th day of December A D 1892

H Selden
Notary Public

Furst, Neu and Company letterhead with drawing of its red sandstone quarry at Portage Entry. (Courtesy Michigan Technological University Archives)

J. W. Wyckoff (1845–1934), Brubaker and Whitesides, photographers, Marquette, photograph date unknown. (Courtesy Marquette County Historical Society)

locally the consolidated Portage Entry Quarries Company until it ceased operations around 1909.

The Portage Entry Quarries Company employed sound business practices. The manager of the company in Chicago, the agent in Marquette, and the superintendent of the quarry at Portage Entry corresponded daily. They discussed stripping new beds; shipping arrangements, including locating scows capable of transporting stone to various ports, the cost of freight, and negotiations with railroad companies; marketing the stone through the Chicago or Marquette office and by bidding on jobs locally; improving or constructing docks and the costs involved; moving stone from one agent or

Quarriers. Portage Entry quarries, Portage Entry Quarries Company, photograph c. 1895. (Courtesy Michigan Technological University Archives, Adolph F. Isler)

yard to another to fill orders; wages and expenses; keeping good workers; maintaining machinery and equipment; renewing contracts and arranging visits; and cutting stone at the sawmill in Marquette.

The quarry on Lot One of Section Nineteen produced a total of 2,537,688 cubic feet of top grade and variegated stone between 1883 and 1898. Production consistently exceeded 300,000 cubic feet annually for the period 1887 to 1891 and peaked at about 450,000 cubic feet in 1890. The quarries on Section Eighteen produced 487,029 cubic feet of stone between 1893 and 1898, averaging 81,171 cubic feet per year. Over time the Portage Entry Quarries Company ran a large business, operating quarries on Sections Thirteen (Township Fifty-three North, Range Thirty-three West), Eighteen, and Nineteen, and on land leased from the Traverse Bay Red Stone Company in Keweenaw County. As Wyckoff explained, the Portage Entry Quarries Company was to commercial red stone what the Calumet and Hecla Mining Company was to copper.[48]

Derricks with workers' houses in distance. Portage Entry quarries, Portage Entry Quarries Company, photograph c. 1895. (Courtesy Michigan Technological University Archives, Adolph F. Isler)

From its operation of the Traverse Bay Quarry in Section Six under lease from the Charles Hebard Estate, the Portage Entry Quarries Company supplied rubble and sawed stone to local mining companies, builders, suppliers, and projects. It furnished rubble stone both to the Baltic, Ahmeek, Allouez, Champion, Centennial, Mohawk, and Trimountain Mining Companies and to such local builders as Paul Roehm, Edward Ulseth, and Peter Donahue.[49]

Charles Hebard of Pequaming had formed the Traverse Bay Red Stone Company in 1894–95 to quarry a pinkish red stone with white mottling in Keweenaw County at the headwaters of the Trap Rock River north of Lake Linden on a 120-acre parcel of land in Section Six, Township Fifty-six North, Range Thirty-one West. Hebard laid eight miles of railroad track from the quarry southeast to land he owned on Keweenaw Bay, where he built a one-thousand-foot dock. He developed

Quarriers. Kerber-Jacobs quarry, Red Rock, photograph 23 July 1895. Kerber-Jacobs Redstone Company. (Courtesy Michigan Technological University Archives, Adolph F. Isler)

and equipped the quarry and extracted sixty-five hundred cubic feet of stone in 1895. The quarry closed in 1896 after twenty thousand cubic feet of stone was produced.[50]

In the meantime, in October 1892, after selling out his interest in Furst, Jacobs and Company, John H. Jacobs assembled a group of investors, including E. H. Tower of Marquette and S. W. Goodale of Detroit, organized the Kerber-Jacobs Redstone Company, and made plans to develop its quarry one mile north of Portage Entry on a 424-acre site on Keweenaw Bay. The

Hewn log workers' houses at Kerber-Jacobs quarry, photograph c. 1895. (Courtesy Michigan Technological University Archives)

investors included a Chicago man named Kerber, probably the son of Henry Kerber, the Chicago cut-stone contractor and partner of Henry Furst from 1861 to 1865, and the Marquette contractors, Powell and Mitchell.[51] Jacobs was president and general manager, and W. J. Fales was superintendent. Before developing the quarry, Jacobs conducted drilling tests for one year to determine the extent and the quality of the sandstone bed.

Once the quality of the stone on the Kerber-Jacobs site was established, Powell and Mitchell employed seventy men to strip the overburden and construct a shipping dock on the unprotected shore of Keweenaw Bay. They opened the quarry one hundred feet from the edge of a fifty-two-foot bluff that forms the shore of the lake in this area. From the opening of the quarry to the lake, workers cut and graded a road in direct line with the dock and on it laid a track over which steam-powered dump cars would haul stone. Parallel to the shore they put in a bulkhead of cribs made from heavy hemlock logs and filled with riprap discarded when stripping the quarry opening. They constructed a dock that extended one thousand feet into the lake.

Along the broad avenue leading from the quarry to the dock grew the settlement of Red Rock. The company's structures stood nearest the quarry. Higher up the hill workers built hewn-log houses side by side. W. J. Fales, superintendent of the quarry, oversaw the construction of two large boardinghouses, a blacksmith shop, barn, storehouses, and offices. Soon the little village of Red Rock took shape in the forest, and the huge scale and full development of the Kerber-Jacobs quarry seemed remarkable. The *Marquette Mining Journal* noted: "The Kerber-Jacobs Company is starting out on a scale never before approached here its equipment being perfect in every feature. It will spring into the field full armed both in quarrying and shipping facilities and as it is believed to have the thickest solid bed of stone ever tackled in this part of the country its operations are already attracting attention and it promises to be a large shipper this year."[52]

In the fall of 1893, Kerber-Jacobs extracted 16,876 cubic feet of stone. By 1895 Kerber-Jacobs shipped 137,529 cubic feet of stone and in 1896, 125,000 cubic feet. It was then the second largest producer in the Keweenaw Bay area.[53] Its success was attributable to the rich bed of stone, the business and technical skill of its president, solid financial backing, and its skillful Finnish-American workers.[54] But by 1895 the sandstone industry at Portage Entry experienced the first tremors of its decline.

Both J. W. Wyckoff and John H. Jacobs assisted other quarry operations. As superintendent of the Furst-Jacobs quarry and on location at Portage Entry, Wyckoff took charge of explorations for the Building Stone and Mineral Exploring Company, an organization of Marquette men exploring for sandstone on land near Portage Entry. Wyckoff presented to the company a detailed technical report showing the results of drill samples to prove that the top grade stone extended in greater density in the property surveyed.[55] Jacobs permitted the Excelsior Redstone Company to transport its sandstone over three miles of railroad from its quarry in Section Eight (Township Fifty-three North, Range Thirty-two West) to Red Rock, then ship it from the docks of the Kerber-Jacobs Company.

The Portage Entry Quarries Company continued the work at the Flag River Brown Stone Company operation in Bayfield County, Wisconsin, until the early 1900s. Incorporated in 1889, the Flag River Brown Stone Company had quarried stone thirty-two miles east of Duluth near Port Wing at Flag River in Section Nineteen. The site was leased from Isaac H. Wing of Bayfield. In 1891 the company extracted over one hundred thousand cubic feet of stone. It employed nearly forty men in its yards at Duluth and fifty at the quarries. When its lease on the quarry expired in 1894, the Flag River Brown Stone Company ceased operation, but the Portage Entry Quarries Company took over.[56]

In April 1908, Joseph Thomlinson, president of the Portage Entry Quarries Company, instructed Wyckoff to limit quarrying that year to fifty thousand cubic feet. Thomlinson explained, "The prospects and outlook for the consumption of red stone is simply deplorable, and all we can depend on is simply additions to present existing buildings."[57]

The Board of Directors of the Portage Entry Quarries Company met on 12 January 1909 and decided not to renew Wyckoff's contract. It notified Wyckoff immediately and explained simply the reason: "The outlook of the colored stone business does not warrant the quarrying of any Portage Red Stone."[58]

Chequamegon Bay

The sandstone of the Bayfield group lies in a six- to eight-mile-wide strip of land bordering Lake Superior in northern Wisconsin. This belt extends almost from the Montreal River on the east to the Saint Louis River, which forms the boundary between Wisconsin and Minnesota at the northwestern corner of the state, and southwestward into Minnesota. Except for bold sandstone cliffs that project at many points along the south shore of the lake and on the Apostle Islands, the sandstone is covered with a thin layer of clay. In Bayfield County, the stone forms nearly vertical cliffs along the lakeshore, some rising forty or fifty feet. In Douglas County, it is found in the channels and banks of nearly all the streams for a distance of one to four miles from the lake and extending to the crystalline rocks. Between 1868 and 1910, twenty-four companies extracted sandstone from fifteen quarries in the Bayfield group. Nine quarries were clustered in the vicinity of Bayfield and Washburn and on three of the Apostle Islands in the Chequamegon formation; six were scattered along the south shore of Lake Superior in the Orienta formation on the Cranberry, Flag, Iron and Amnicon Rivers. One company extracted stone from a quarry on the Saint Louis River at Fond du Lac, Wisconsin.

F. T. Thwaites, assistant state geologist in Wisconsin, identified the quarry sites producing good stone in the Chequamegon and Orienta formations:

> It will be noted that the quarries, especially those in the Apostle Islands, are arranged in a nearly straight line along the strike of the formation. Therefore, but one horizon of good quarry rock is believed to exist in the Chequamegon formation, although good stone was seen at many scattered points. Less is known about the Orienta sandstone. The best brownstone is that found at Port Wing and Orienta, but it is most probably that the brownstone quarry on Amnicon River is at a the lower horizon. The quarry at Fond du Lac does not produce a true brownstone, the

rock being firmer, and more felspathic, and irregularly colored red and white.[59]

The sandstones of the Bayfield group vary in color from light pinkish reddish brown to dark bluish purplish brown. Some are variegated with white streaks and spots. The *St. Paul Pioneer Press* described simply and romantically the sandstones from the Duluth Brownstone Company's quarry at Fond du Lac and the Bayfield Brownstone Company's quarry on Chequamegon Bay: "They are essentially the same stones, both being Potsdam sandstones, colored to their beautiful brown by some antediluvian alchemy which diffused through every particle of rock its necessary atoms of iron pigment."[60]

In the Chequamegon Bay area, companies opened quarries on Basswood (Bass), Stockton (Presque Isle), and Hermit (Wilson) Islands, and at Houghton and Van Tassell's Points. Production was heaviest between 1883 and 1896. It peaked in the early 1890s when the Ashland Brown Stone Company, the Excelsior Brownstone Company, and the Superior Brownstone Company quarried stone at Stockton Island, Hermit Island, and Basswood Island, respectively; when the Washburn Stone Company, the Hartley Brothers, and the Prentice Brownstone Company quarried stone at Houghton Point; and when Robinson D. Pike extracted stone from Van Tassell's Point. The *Ashland Daily Press* reported the total production for 1892 for the seven quarries in the Chequamegon Bay area at 2,313,000 cubic feet.[61]

The companies operating in the Orienta formation did not flourish to the same extent as those working in the Chequamegon formation because, in part, they lacked good harbors for water transportation.

At Fond du Lac, red rocks in Wisconsin are assigned to the Orienta formation; those in Minnesota are assigned to the Fond du Lac formation. The two are probably correlated. However, there is an intervening gap of twenty miles between definite Orienta and definite Fond du Lac, and proof of correlation is not possible.[62]

Frederick Prentice and the Prentice Brownstone Company Quarry

The largest and most ambitious quarry operation in the Chequamegon Bay area during the period when the burgeoning industry grew to immense size was the Prentice Brownstone Company. The company owned one mile of water frontage three miles south of Washburn at Houghton Point in Section Twenty-seven, Township Forty-nine North, Range Four West.

In 1888 Frederick Prentice and Ashland and New York City investors organized a corporation with capital stock of $1,250,000 to develop a quarry on land that Prentice had owned for thirty years.[63] Incorporators were Edwin Ellis, Eugene A. Shores, Cassius M. Hamilton, and George H. Barr.

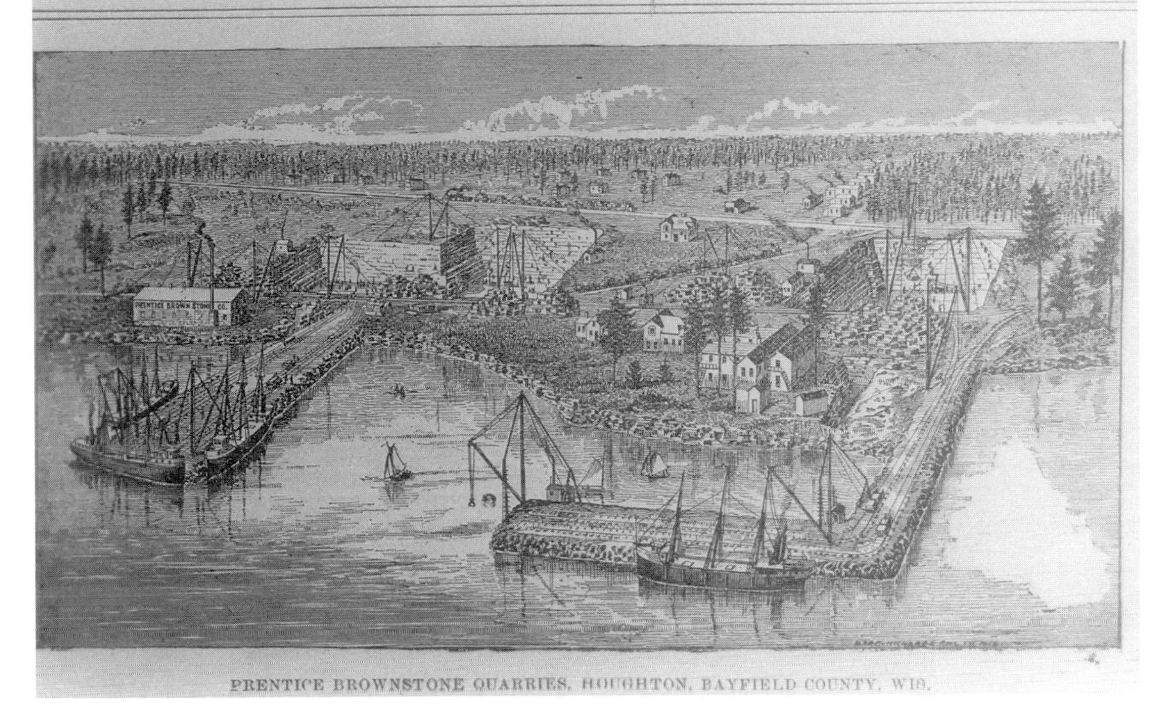

Prentice Brownstone quarries at Houghton, Bayfield County, Wisconsin. From Ashland Daily Press, *Annual Edition (1893) 26.*

The previous summer a large work force cleared land and stripped earth from the surface for what the local newspaper forecast would be one of the largest stone quarries in this section of the country. The Chicago, Saint Paul, Minneapolis and Omaha Railroad laid a side track right up to the works, and tramways led from the quarry to docks so stone could be shipped either by rail or by ship. Within one and two years, the company employed between 200 and 250 workers, and the quarry town of Houghton had a Main Street, numerous workers' cottages, three boardinghouses, sawmills, a general supply store, and a blacksmith shop. The operation itself employed eleven steam channelers, twelve derricks, and a large sawmill.

Prentice said that he intended to ship the stone chiefly to New York City by way of Buffalo and the Erie Canal. He anticipated his principal markets to be New York, Boston, and Philadelphia, but thought he would also ship to Chicago, Milwaukee, and Detroit. Prentice sold the stone locally as well.

Two years after opening the quarry, Prentice installed a large sawmill and ran it year round. He erected a combined storehouse, workshop, stables, and barn, and installed a reservoir tank and water system. He ordered a stone traveler—a motorized machine fitted to run on a track above the quarry that took out stone and piled it up for shipment—and a self-propelled stone piling car capable of drawing two or three loaded cars. Both enabled the company to stockpile a large quantity of stone through the winter for spring shipment. Explained Prentice: "We are not trying so much to get out a large quantity of stone, as to get the quarries in thorough condition for handling orders of almost any magnitude in quick time."[64] The quarry yielded 383,887 cubic feet of stone in 1889 and 623,334 cubic feet in 1890.

His operation at Houghton Point underway, Prentice expanded his scope to the Apostle Islands. In August 1890, E. E. Davis of Ashland began prospecting for the location of a quarry on Hermit Island. Prentice purchased for eighty-five hundred dollars the Hermit Island land in the southeast one-quarter of Section Thirteen from Elias F. Drake of Saint Paul in October 1890. Drake had bought it for thirty-five hundred dollars two years earlier from Julius Austrian of Saint Paul and members of the Aaron Leopold Family of Chicago. Leopold and Austrian, frequent speculators in Lake Superior region land, in turn, had acquired their interest from earlier speculators in 1870 and 1885, respectively.

Frederick Prentice began operating the quarry on Hermit Island on 2 May 1891. Within six weeks, the first shipment of stone left Hermit Island bound for Buffalo. The first year three vessel cargoes and five barges of stone were taken from the island.

Initially, Prentice spent $80,000 on improvements at the quarry, and within a short time he had spent $150,000 in all. Prentice had purchased Cook and Hyde's real and personal property on the island and the mainland in March 1889. The machinery and equipment from this acquisition no doubt were installed at both the Houghton Point and Hermit Island quarries. Machinery on Hermit Island included an engine, channelers, and steam derricks. Prentice employed one hundred men on Hermit Island and built a village of cottages. On a high promontory near the quarry, he erected a rustic summer cottage veneered with cedar bark for himself and his family.

Their lease expired in 1888, and Cook and Hyde agreed to the sale to Prentice because, as they explained on 30 March 1889, "We only went into the quarry business to furnish first-class stone for our yards in Milwaukee and Minneapolis, and your quarries are so well prepared to furnish first-class stone on short notice, we have decided to accept your proposition to purchase our quarries on the mainland and islands, with our channelers, steam derricks, powers, and all machinery at both quarries, and take stone of your company, and will make the deeds for the property out at once."[65]

In 1883 Cook and Hyde had leased the Basswood Island quarry from its Milwaukee and Chicago owners and had operated it from May of that year until March 1888 to supply stone for their yards in Milwaukee and Minneapolis. For forty years, from 1858 to 1898, Edwin Hyde (1828–1909), in partnership with Thomas D. Cook, dealt in cut stone in Milwaukee and Minneapolis and had a contracting business in Milwaukee. Cook and Hyde were among the largest contractors, builders, and consumers of Lake Superior red sandstone in the Northwest. To supply their market, the firm quarried stone on Basswood Island from 1883 to 1888 and on the mainland in Bayfield County at a sixty-acre site adjacent to the Bayfield Brownstone Company quarry from 1886 to 1889. They purchased ninety-six acres on Hermit Island in 1884 with the intention of opening a quarry there.

Cook and Hyde leased Lot One in Section Four (Township Fifty North, Range Three West) on Basswood Island from Robert Strong, Daniel Wells, Edwin French, George Lee, and Edwin Walker, who still held title to the land after a lawsuit. Lot Two may have been leased from Beriah Magoffin. Cook and Hyde made some capitol improvements at the quarry. In December 1883 workmen built a long, low rambling log structure, probably a boardinghouse for workers. In the summer of 1884, they extended the dock to accommodate boats with twelve-foot drafts and installed two new steam drills and one steam derrick.

F. C. Bailey of Milwaukee managed the quarry operation on Basswood Island for Cook and Hyde from 1883 through 1885. Each year Bailey arrived on the island in April and stayed through November. Thomas Cook and Edwin Hyde themselves made periodic visits to check on activities at the quarry. While in the area, they prospected for new quarry sites and bid on local construction jobs.

By 1883 the completion of the Chicago, Saint Paul, Minneapolis and Omaha Railroad to Bayfield made possible shipments by rail as well as by water. Although Cook and Hyde continued to ship most of the stone by water in six-hundred-ton loads to Milwaukee, they also transported stone inland by rail to Madison, Wisconsin, Minneapolis, Iowa, and elsewhere. The firm probably shipped more tonnage in 1884 than in any other year.

Cook and Hyde furnished stone for Saint Paul's Episcopal Church (1882–91), T. A. Chapman's Dry Goods Store (1884–85), and the Plankinton House Hotel (1885–86) in Milwaukee; the Germania Bank Building [Saint Paul Building] (1888–90) in Saint Paul; and the Bayfield County Courthouse (1883–84) in Bayfield.

The Basswood Island quarry site, operated by Strong, French and Company in the early 1870s but idle for nearly ten years following, had supplied stone for the Milwaukee County Courthouse in the early 1870s. Cook and Hyde became interested in the Basswood Island quarry site in July 1882. The building committee of the vestry of Saint Paul's Episcopal Church in Milwaukee had selected the plans for the church prepared by

Edward Townsend Mix in May 1882, and things began to happen. Edwin Hyde, a cut-stone merchant and contractor, traveled with Mix to Basswood Island to complete arrangements for shipping brownstone to Milwaukee. The need for stone for Saint Paul's Church and the upturn in the economy in the early 1880s created a demand for Lake Superior sandstone. Once on Basswood Island, Mix and Hyde dealt with Joseph McCloud, an island resident who acted as caretaker of the quarry and agent for Milwaukee and Chicago owners, after Strong, French and Company had ceased quarrying in 1873.

To begin to fill the order for Saint Paul's Church, McCloud probably turned to stockpiles of sandstone left from Strong, French and Company's operations. And he may also have sent to Milwaukee stone extracted by an unnamed firm that reportedly leased the site in 1879 and removed a fair amount of stone for a few years. Some of the stone for the church may have even come from Sand Island, for the *Bayfield Press* for 5 August 1882 reported that a Milwaukee vessel had taken on stone for "the new Episcopal Church there" from a quarry on Sand Island.[66] Or, perhaps a Mr. Maxwell, who, according to a *Bayfield Press* account of 26 August 1882, came to Basswood Island with a crew from Cleveland and worked the quarry until the winter freeze, may have supplied some of the stone for Saint Paul's Church.

By 1893 the Prentice Brownstone Company neared bankruptcy. To raise money to cover debts and to meet operational expenses, Prentice reorganized the corporation and conveyed his real estate holdings to the company. Thus, on 1 June 1893, Prentice and his wife conveyed to the Excelsior Brownstone Company lands that included all of Hermit Island except the 50-acre site of his cedar-bark summer cottage and surrounding park, 304 acres on Stockton Island, 227 acres on Hemlock Island, and nine lots in Ashland containing the large stone dock.

Three weeks later, S. S. Fifield, Edwin Ellis, F. E. Goddard, and L. C. Tobias filed articles of incorporation for the Excelsior Brownstone Company of Ashland. The capital stock was $1.8 million. The next day the Excelsior Brownstone Company authorized the Board of Trustees to issue bonds in the amount of three hundred thousand dollars for the purchase and improvement of company lands and for the carrying out of company purposes. The bonds were backed by a mortgage of the lands Prentice had recently conveyed to the Excelsior Brownstone Company. Also included in the sale were steam tugs, a scow, sailboats, tools, machinery, equipment, a team of horses and wagons, and brownstone lying on the docks.

Production at the Excelsior Brownstone Company quarry reached 220,000 cubic feet of stone per year for the period 1893 to 1895. By 1897 operations had ceased. Frederick Prentice and the Excelsior Brownstone Company relinquished title to all of Hermit Island to the estate of Elias

Drake by default on mortgage payments. The land was sold at a referee's auction for ten thousand dollars. In 1902 W. G. Maginnes of New York City purchased for $7,250 the Prentice quarries on the mainland and Hermit Island, which had been held in receivership for ten years.

From the time it opened until it went bankrupt in 1893, the Prentice Brownstone Company conducted business on a large scale. It furnished stone for the Cincinnati City Hall (1888–93), the Lumber Exchange (1885) in Minneapolis, the Potter Palmer House (1889) in Chicago, the Washington High School in Saint Paul, and the Ashland Post Office (1892–94) in Ashland.[67]

Typical of Prentice's expansiveness was his scheme to extract a monolith obelisk from the quarry and exhibit it at the 1893 World's Columbian Exposition in Chicago. The monolith was to be a monument to Ashland brownstone as well as a monument to himself. Measuring 113 feet in height, exceeding that of the quarried Egyptian obelisk, this solid brownstone monolith was first planned to be taken from the Excelsior Brownstone Company quarry but was, in fact, extracted from the Prentice quarry and readied for shipment. The project was never brought to completion because of the lack of funds to ship and install the awesome monument. However, four twenty-five-foot monoliths, a life-size statue of a Chippewa chief, a statue of a Wisconsin badger, and heads of African Americans in bas-reliefs—all carved in brownstone—were shipped by Prentice to the World's Fair. The *Ashland Press* for 31 December 1892 proudly proclaimed that for the Ashland fair goers "the familiar sight of brownstone will greet them everywhere."

Summary

Companies extracted Jacobsville and Bayfield sandstone from sites in the Lake Superior region from before 1870 until the early 1900s. They shipped the stone easily and cheaply by water and by rail to the Midwest's important commercial, trade, and industrial centers. The stone was also used locally. The years of operation followed economic cycles. Activity rose in the early 1870s, 1880s, and 1890s, but fell off with the economic declines of 1873, 1893, and 1897. The period of greatest activity took place from 1883 until 1896. Then companies quarried sandstone at Marquette, Portage Entry, and Houghton Point as well as on the south shore of the lake and on the Apostle Islands.

Quarry companies and quarry lands were owned, developed, and managed by investors in the large midwestern cities of Detroit, Milwaukee, Chicago, and Saint Paul, and in the local cities of Marquette, Houghton, Ashland, and Superior. Many owners of quarry companies were stone merchants and contractors in their resident cities. Furst, Neu and Company;

Strong, French and Company; and Ashland Brown Stone Company had stone yards in Chicago. Cook and Hyde had stone yards and a contracting business in Milwaukee and Minneapolis. The Marquette Brownstone Company had stone yards at Marquette, and the Superior Brownstone Company at Superior. Companies without stone yards quickly established them. Frederick Prentice opened stone yards in Chicago and New York City, and Knight opened stone yards in Ashland. Some companies extracted stone from more than one location.

The sandstone was plentiful, but men to operate the quarries and funds to prepare the quarry site and purchase and install machinery and equipment were not. Companies brought in crews and foremen from their resident cities. The constant need for money for development and capital improvements required directors to gain the approval of stockholders to restructure and reorganize to increase capital stock, to mortgage assets and sell bonds, or to sell out to other investors.

Quarry production paralleled economic and social circumstances in the cities in which the stone was marketed. People and cities needed to replace frontier structures with solid and substantial buildings appropriate to their new economic and social stature. Lake Superior sandstone furnished a beautiful, strong, and durable material with which to build them.

At the moment the sandstone quarry industry rose to its period of greatest production, and the cities and villages of the Lake Superior region reached their period of greatest prosperity and growth, the massive, ponderous, stone forms of the Richardsonian Romanesque architecture came into popularity in the Midwest. Named for Henry Hobson Richardson (1838–86), this mode was particularly suited for expression in colored stone. Richardson had given name to the style through his designs for buildings that borrowed elements from the medieval French Romanesque and massed towers, arches, and buttresses in varied colors and textures of Connecticut brownstone and Hallowell granite. The stone and the style gave the builders of the Lake Superior region an architectural medium capable of expressing their ambitions, tastes, and fears, durable enough to withstand a harsh climate, and rugged enough to harmonize with wild rocky surroundings.

The quarry industry declined in the late 1890s. After the depression of 1893, employment and production at the quarries fell off since supplies on hand were sufficient to fill orders, and many quarries became idle. The value of sandstone produced in Michigan plunged from $290,578 in 1901 and $188,000 in 1902 to $13,000 in 1911.[68] The director of the Michigan geological and biological survey cited as causes the depression of the 1890s and the great distance of the deposits from large markets. Also, lighter-colored natural stones had gained favor. Then, too, artificial stone, concrete, and brick displaced natural stone. These man-made materials cost less and could be handled more rapidly and easily and conformed to contemporary taste in architecture.[69] Moreover, brick manufacturers mounted a successful campaign against stone.

The commissioner on mineral statistics reported 1895 an unprofitable year for producers of Lake Superior sandstone and explained the changing fashion: "Architects have pronounced against it in their plans and specifications, not because it is not substantial or beautiful, but for the reason that architecture must have change of style and material the same as millinery and tailoring. They claim that too much sandstone was being used, and that the sameness must be broken into by the use of stone of other kind and color."[70]

The decline of the fashion for Lake Superior brown sandstone greatly disappointed and perplexed many residents of the Lake Superior region. On learning that Portage Entry red sandstone had lost out to Bedford white limestone for the construction of the Federal Building at Sault Sainte Marie, despite the opposition of the Business Men's Club and prominent Upper Peninsula citizens to federal architect John Knox Taylor's decision, the agent of the Portage Entry Quarries Company in Marquette observed, "So here we have a building, next door to the quarry, so to speak, requiring about 16,000 feet gone into Bedford."[71]

By then, the quarries had yielded beautiful, durable, and carvable red sandstone that met the needs of architects, builders, and clients constructing ever more substantial architecture in the cities and villages across the Lake Superior region.

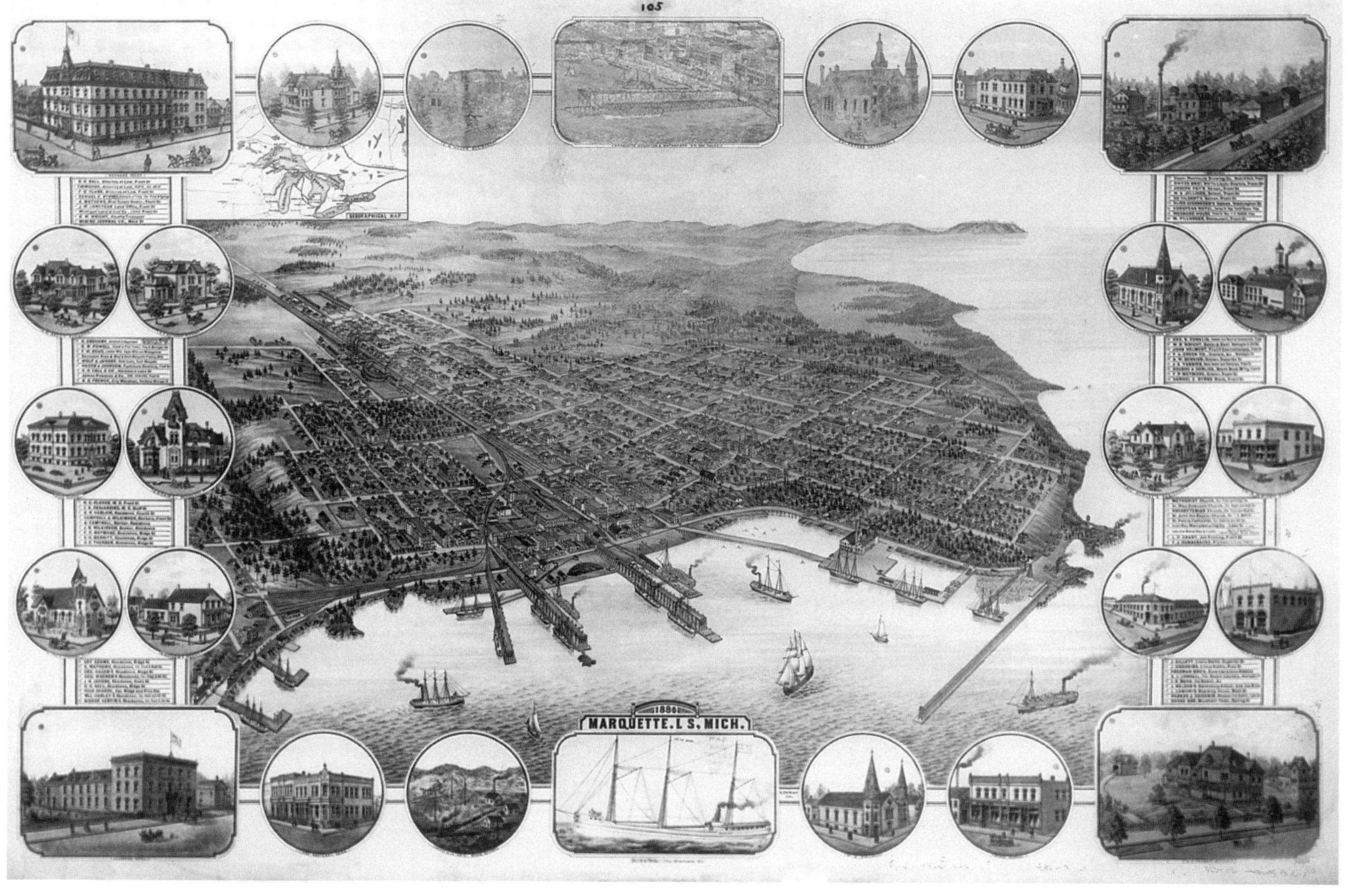

Marquette. L.S. Mich., 1886. Beck & Pauli. Litho., Milwaukee, Wis. Showing clockwise vignettes of the Marquette brownstone Saint Peter's Cathedral, Methodist Episcopal Church, First National Bank, Saint Paul's Church, High School and Sidney Adams House. Also shown bottom center a vignette of Furst, Jacobs Stone Quarry. (Courtesy Marquette County Historical Society)

2

THE SANDSTONE ARCHITECTURE OF MARQUETTE

INTRODUCTION

Marquette is located on Marquette Bay, an inlet of Lake Superior midway between the Saint Marys and Montreal Rivers, which form the eastern and western boundaries of the Upper Peninsula. The city nestles beneath highlands that rise first to a plateau, then to a chain of hills. Belts of geologic formations encircle the community, and rocks crop out everywhere. The land all around is covered with heavy forests of pine and hardwood and broken by rivers and valleys.

The iron in the hills around Marquette was known to the Indians and French missionaries of the early seventeenth century and to trappers of the early nineteenth century, but it was not until 1844, when William Burt and Jacob Houghton (1827–1903), the brother of Douglass Houghton, discovered iron deposits near Teal Lake, twelve miles west of Marquette, that development of the region began. Their discovery substantiated Douglass Houghton's earlier reports of extensive iron deposits and this, together with subsequent findings, brought an onslaught of entrepreneurs and fortune hunters to recently organized Marquette County.

Following Burt and Houghton, Philo M. Everett (1804–92) explored the area in 1845 and located lands for the newly formed Jackson Mining Company, the first organized mining company in the region. In 1846, after sending one ton of ore to the mouth of the Carp River and then on to Pittsburgh for scientific testing, he opened the Jackson Mine. A year later, three miles east of Negaunee on the Carp River, Everett built the Jackson Forge for the production of iron.

In 1849 Amos Rogers Harlow, Waterman A. Fisher, and Edward Clark, all of Worcester, Massachusetts, together with Robert J. Graveraet of Mackinac Island, organized the region's second iron concern, a forge known as the Marquette Iron Company. On learning of the discovery of iron ore on Lake Superior, Harlow consulted with J. D. Whitney of Boston, who with John Wells Foster had conducted the geological survey

Ore Docks, Marquette Harbor, photograph date unknown. The modern ore dock at Marquette's excellent natural harbor permitted the efficient shipping of iron ore from the mines of the Marquette Iron Range to the coal fields of the lower lakes. (Courtesy Library of Congress, Detroit Publishing Company Photographic Collection)

of the Upper Peninsula. Encouraged by Whitney's report, Harlow and his party moved forward and later that same year reached the present site of Marquette. Attracted by the area's excellent harbor, they cleared ground, erected a few simple wooden buildings, and constructed a forge for the production of iron from ore that could be transported from nearby mines.

By 1853 three mining companies—the Jackson, the Lake Superior, and the Cleveland—operated in the Marquette Iron Range, but further development required an effective means of transporting the ore from the mines to the large furnaces that were located near the coal fields of the lower lakes. The completion in the 1850s of the first railroad and modern ore docks at Marquette and Escanaba and the opening of the locks at Sault Sainte Marie provided such facilities. Mines at Ishpeming and Negaunee followed, with others at Gwinn, Republic, Champion, and Michigamme established soon after. A network of railroads connected the mines to the docks at Marquette and Escanaba.

As the rail network expanded later in the century, Marquette became the Upper Peninsula's leading shipping center. Prior to the advent of rail transportation, Marquette bore little resemblance to the flourishing "Queen City of the North" characterized by later writers. By 1862, however, the community's population exceeded sixteen hundred, and investment returns on mining and shipping interests sent the local economy soaring. Lumbering, the extraction of sandstone, and tourism followed. The city fast became a regional center of commerce, finance, government, and speculation in land, minerals, and timber.

Building activity created demands for materials and labor that the people of Marquette could not meet, despite the region's vast resources. Aware of the shortages, architects, carpenters, builders, and suppliers in Detroit and Cleveland advertised their goods and services in the Marquette newspapers for the 1860s. In their quest for high-style results, Marquette clients in turn called upon architects from Chicago, Cleveland, Milwaukee, and Detroit to design their increasingly complex buildings. By 1875, however, the city was able to claim its own building suppliers, sandstone quarries, builders, carpenters, and masons; and for three years, even an architect, the Norwegian-born Gothic Revivalist, Carl F. Struck, was in residence. Finally, around 1890, Demetrius Frederick Charlton, a fully trained architect, opened a practice in Marquette.

From the time of settlement, building stone had been blasted out of quarries and taken from excavations to be used for foundations, mines, and roads. It was not until industry developed a need for substantial and utilitarian structures and brought into the region the engineers, workmen, and masons capable of understanding and handling stone, that it was used for building purposes. Once it became available, however, mining, furnace, and railroad companies, as well as government agencies, found the local sandstone ideally suited for varied building needs. Between 1869 and 1900, more than a dozen companies intermittently extracted brown sandstone from quarries in Marquette County. What was not used locally was shipped to large midwestern cities.

One of the early stone masons in the Upper Peninsula was English-born George Craig, who arrived in Marquette in 1860. Under the supervision of noted iron man Stephen R. Gay, Craig erected the Forestville Furnace (1860s) on the Dead River and Greenwood Furnace (1867) at Morgan. He also built the Marquette and Pacific rolling mill. Then, working for master mechanic Jeremiah P. Van Iderstine, superintendent of construction for the Iron Mountain Railroad, Craig constructed cuts through the rocks leading to the docks. Craig probably extracted some of the first stone for the Samuel Peck Building. The stone came from a partially opened quarry on the former J. P. Pendill Farm, which he and a man named Smith owned in the southern portion of Marquette. Samuel Peck himself was among the first to dress the sandstone from Marquette quarries for building purposes. Stone from this quarry was used for several business blocks and for the roundhouse and machine shops that Craig later built for the Marquette and Ontonagon Railroad.[1]

The Peter Wolf and Son firm of Chicago, which had purchased the Smith and Craig property in 1870, shipped almost all of the large blocks to Chicago and cheaply sold the rubble locally for use in cellar walls and foundations.

Rubble and crudely finished stone blasted from quarry sites in Alger and Marquette Counties was used in such industrial structures as the mighty Bay and Schoolcraft Blast Furnaces on Munising Bay and in the little worker's cottage near an early South Marquette quarry.

Bay Furnace, Onota

In 1869–70 the Bay Furnace Company built a blast furnace at Onoto, situated on Grand Island Harbor on the south shore of Lake Superior six miles west of Munising and forty miles east of Marquette. Workers blasted light bluish red rock from nearby Powell's Point and used it to erect a stack large enough to accommodate both the hot blast and boilers that were placed at the top. They gathered stone on the spot and built two sets of charcoal kilns two and one-half miles from the furnace and two and one-half miles from each other. Each set was located in the center of great tracts of hardwood timber and produced the charcoal fuel that was hauled over roads to the furnace. The furnace fired iron ore, charcoal, and flux in a continuous process and produced pig or cast iron. Ships carrying iron ore from Marquette and pig iron to the steel mills of the lower lakes unloaded and loaded at a nearby company-owned dock. The Bay Furnace, focal point of Onotoa, began smelting iron in the spring of 1870 and produced 3,498 tons of pig iron that year. By 1874 the furnace turned out 8,359 tons and employed six hundred, including those engaged in making charcoal. Fire destroyed the Bay Furnace in June 1877, and only stabilized ruins remain.[2] Today Bay Furnace is in the Hiawatha National Forest.

Schoolcraft Furnace, Munising

Similarly, the Schoolcraft Iron Company selected and extracted stone for its furnace on Grand Island Harbor at Munising. After finding the rock blasted out for the foundation too soft for building purposes, the company opened a quarry on the southwest end of Grand Island and transported the stone across the bay by scows. The *Marquette Lake Superior Mining Journal* for 21 November 1868 praised the result as a handsome and substantial piece of masonry. Two Marquette residents, Peter White and

Henry R. Mather, formed the Schoolcraft Iron Company in 1866. The furnace was located along Munising Creek some eleven hundred feet from the shore of Lake Superior where it had a natural harbor and a hardwood supply. In 1870 the furnace produced sixteen tons of pig iron a day.[3]

John Burt House

The John Burt House at 220 Craig Street in Marquette further exemplifies early rubble stone construction in Marquette. The house is the oldest surviving structure of its type in the city. Built in about 1860 by John Burt, the stone cottage subsequently served as a warehouse and clerk's office for the Burt Freestone Company established in 1872 to work a nearby quarry. Two-feet-thick rubble stone walls, later resurfaced with a stucco-like material, rise of the snug little one-story cottage. The gable roof presents its long side to the street, and four small windows admit light to the attic space. The interior is divided into two rooms, one of which contains an open fireplace.

By the 1870s Marquette had progressed socially and economically to the point where the residents were ready to use the native reddish brown dimension stone for their own domestic and public buildings. The need for substantial new structures, some to replace primitive wooden ones, and the disastrous fire of 11 June 1868 that destroyed both the business district and much of the residential section, hastened the need to rebuild.

The fire started in the Marquette and Ontonagon Railroad shops and destroyed all the business blocks then in existence north of Superior Street; the Marquette and Ontonagon Railroad shops themselves, which then occupied the north half of the block between Front, Third, and Spring Streets; the Marquette and Ontonagon dock, warehouses, and trestle dock; and the Lake Superior Company's pier.

Although it took two years to rebuild the vast sections damaged by the fire, five months after it happened the *Marquette Lake Superior Mining Journal* concluded that the fire would benefit both the business and the physical appearance of Marquette. Many businessmen burned out in the fire would build more elegant and commodious structures. The newspaper observed that the substantial new stone and brick buildings would represent greater value than all the wooden structures destroyed by the fire.[4] A Wisconsin writer described the shift from building with wood to building with brick and with stone in these comments about Marquette:

> Fine buildings of stone and brick have taken the place of the cheaper wooden structures, and have given [Marquette's] business center the more substantial and lasting look of a New England town. . . . Among the promising producing interests of Marquette, most recently devel-

> oped, are the quarries of brown freestone, which are now being worked to a considerable extent, and the product sent to Milwaukee, Chicago and other cities for building purposes. It is the same stone, we presume, so popular in New York in the brown stone fronts of buildings. It comes out of the quarry quite soft, and it is easily worked; but with exposure to the air soon hardens. The supply is inexhaustible, and the hewn stone will doubtless become one of the most valuable products of the place for export. Some buildings have already been built of it in Marquette—as the basement of the new Methodist Church and of the new public schoolhouse.[5]

Marquette was incorporated as a city in 1871, and during the twenty years following the fire, the population tripled. By 1890 its population had reached ten thousand, and despite the panic of 1873, the community experienced one of its biggest building booms. The many fine residences and commercial buildings attested to the city's prosperity.

Marquette and Ontonagon Railroad Company Shops and Roundhouse

The Marquette and Ontonagon Railroad Company built in 1868 a group of machine shops and a roundhouse to replace those destroyed by the fire. To the people of Marquette, this valuable improvement demonstrated the company's faith in the future of the city.[6] The utilitarian structures were constructed of beautiful Marquette brownstone quarried by blasting and roughly finished. Solid, substantial, and fireproof, the roundhouse, machine shop, blacksmith shop, and foundry stood grouped for convenient mutual use and access next to the tracks on company grounds in the western part of Marquette.[7]

The machine shop, blacksmith shop, and foundry were contained in a single long building under one roof. The machine shop held all the tools and equipment—lathes, boring and drilling machines, screw cutters, and hydraulic press—needed for supplying wheels and axles for several hundred ore cars. The tools were arranged parallel to a railroad track that ran along the midline of the structure. The blacksmith shop had one dozen fires, belt heaters, and the usual accessories. Twelve or fifteen men worked in the foundry. The roundhouse contained nine stalls that sheltered locomotives under repair and rebuilding. With stone walls and with iron doors, window frames and casings, the roundhouse was as fireproof as possible.

There were other structures, including a car shop and forge, carpenter and pattern shops, lumber sheds, and storehouses. Once isolated from shops and mills, the company gradually built its shops to manufacture and

Roundhouse of the Marquette and Ontonagon Railroad Company, 1868–70, Marquette, photograph date unknown. Destroyed. (Courtesy Marquette County Historical Society)

maintain its railroad cars, locomotives, tracks, and bridges. By 1870 the Marquette and Ontonagon Railroad employed from 150 to 200 workers at these shops.[8]

Superior Building (Mather Block, First National Bank Building)

Stone first appeared in the business blocks of Marquette in the 1870s. Eight years after the First National Bank had organized on 22 January 1864, prominent Marquette businessmen Henry R. Mather and Peter White together with their partners, the directors of the bank, built the Italianate Superior Building at the southwest corner of Front and Spring Streets, one of the city's principal intersections.[9] The building would set the standard for later commercial blocks like the Marquette County Savings Bank.

As soon as President Lincoln signed into law the National Banking Act, White, Mather's brother-in-law and operator of a private bank since 1853, interested some of Marquette's industrial and business leaders in forming a national bank.[10] To design the city's first major commercial block, these people whose connections and experiences extended beyond Marquette, sought a well-trained urban architect. At the time, Henry Lord Gay (1844–1921) was establishing his reputation in architecture in Chicago.

Gay had arrived in Chicago in 1864 and had worked with W. W. Boyington, a noted hotel architect, until he opened his own practice in 1867.[11] Born in Baltimore, Maryland, Gay attended private grammar school in New Haven, Connecticut, and later trained with Sidney Stone of that city. Gay dispatched his associate, Norwegian-born Carl F. Struck (1842–1912), from his Chicago office to Marquette to supervise construction work contracted to Alfred Green of Cleveland. Struck was born in Oslo, Norway, educated in Copenhagen, and emigrated to the United States in 1864. With their involvement in the Superior Building, Struck and Green announced plans to open offices in Marquette. All three pointed to the Superior Block as evidence of their unique talents.

Gay opened a branch office in Marquette in 1872, placed Struck in charge, and sought commissions for public buildings, churches, schools, hotels, and private residences. Similarly, Green established an architectural design and building business in Marquette and eventually managed the newly organized Lake Superior Building Company.[12] Struck remained in Marquette for three years and set standards of quality for architectural designs during the 1870s.

On 16 September 1871 the *Marquette Mining Journal* announced that Mather and the First National Bank had begun their building and predicted that it would be one of the best blocks in the city. In setting out to build the handsomest and most substantial business structure on Lake Superior, they constructed it of variegated brown Marquette sandstone extracted from the Wolf quarry by plugs and feathers; hand cut, rubbed and polished; and laid in even courses.[13] While the construction of the block was in full progress, White, Mather, and Green, together with other residents of Marquette, formed the Marquette Brownstone Company and bought out the Wolf and Sons business, equipment, and quarry in South Marquette. Construction on the Superior Building was completed sometime in 1873. Estimated at seventy-five thousand dollars, the cost probably exceeded one hundred thousand dollars.

The Superior Building rose three stories from a raised foundation and fronted on Front Street for ninety-six feet and Spring Street for eighty-five feet. Dressed stone was used for the lintels, belt courses, and door jambs as well as for the walls. Wide stairs led to the corner entrance. Round and seg-

Superior Block, 1871–73, Henry Lord Gay and Company, 1885, Hampson Gregory, Marquette, photograph date unknown. Destroyed. Superior Building is on the right. (Courtesy Marquette County Historical Society)

mentally arched windows graced the Front and Spring Streets facades, and ornamental pediments projected from the ample decorative cornice. The words "Superior Building" were carved in the stone front, "Bank" over the corner entrance, and "1872" on the center pediment. The ground floor was arranged with large stores and rooms for the First National Bank. The second floor held offices, and the third floor had a large public hall. Black walnut, white ash, and chestnut richly finished the interior throughout. A basement furnace heated the building.[14]

Twelve years after the completion of the magnificent brownstone Superior Building, the people of Marquette still took pride in its solidity and imposing architectural beauty. The business and commerce of Marquette seemed to revolve around this building.[15] In early February 1885, a fire destroyed the building, then also known either as the Mather Block or as the First National Bank Building.

Mather then commissioned Hampson Gregory (1833–1921), a Marquette builder, to plan the reconstruction in two rather than three stories and to rebuild it during the summer and fall of 1885. Gregory raised the first floor so that the full basement would accommodate offices, apartments, and businesses, placed the bank in the corner of the first floor separated from a store by a stairway to the second floor, and divided the second floor into six offices and an auditorium that seated seven hundred. The replacement of the building cost less than one-half the cost of original construction because Gregory modified the design and used stone and brick salvaged from the original building. During the construction of the original building in 1871–73, the labor required for hand sawing the stone and the freight charges for shipping in materials and supplies over an unfinished railroad system had been very expensive.[16]

High School (Ridge Street School)

Publicly funded utilitarian sandstone structures differed in character from privately funded sandstone structures. Built in 1874–75 but destroyed by fire in February 1900, the high school (Ridge Street School) was the first publicly funded building erected of sandstone in Marquette. Carl F. Struck was the architect, and Hampson Gregory was the builder. Resident practitioners working with a local building committee intent on giving the community the highest value for its investment produced a functional design. The no-nonsense approach to building materials taken by the Marquette School Board when it replaced the brick school that had occupied this site since 1860 clearly indicated its sense of accountability to the taxpayer and its awareness of the value the community placed on education.

In 1874, one year after thirty thousand dollars had been appropriated for the construction of a new high school, members of the Marquette School Board contemplated a building of mottled stone—less expensive than clear uniform stone but considered equally beautiful—trimmed with clear brown stone.[17] The board quickly adopted Struck's plans, and the *Marquette Mining Journal* for 20 June 1874 explained the architect's intent in these words: "There has been no attempt at fancy architecture, the external appearances of the building being at once neat and massive, suggestive of solidity, and devoid of superfluous ornamentation. . . . The arrangement of the interior shows convenience and adaptability to have been the study of the architect, while the solid, substantial appearance of the building externally will be pleasing, and convince the public that their money has not been frittered away by sacrificing stability to a taste for gorgeous effect." To the Marquette School Board the variegated native sandstone seemed the most suitable building material for the utilitarian nature of a public school and for conveying the board's wise use of public monies. Only minor concessions were made to ornamentation.

High School, 1874–75, Carl F. Struck, Marquette, photograph date unknown. Destroyed 1900. (Courtesy Marquette County Historical Society)

From 1873 to 1874, before moving west to LaCrosse, Wisconsin, and to Minneapolis, then to Tacoma, Washington, Struck lived in Marquette, designing beautiful Gothic and Italianate stone houses, schools, and churches.[18] Struck's buildings seemed handsome and substantial, and his role in improving the appearance of Marquette did not go unnoticed.

The Marquette School Board awarded the construction contract for the high school to Hampson Gregory. Born in Devonshire, England, Gregory had studied building while he worked as a miner. In 1860 he emigrated to Canada where he worked in the construction business for eight years. He came to Marquette in 1868, just as the city was rebuilding after the fire. Soon after arriving in Marquette, Gregory advertised as a builder competent in brick, stone, and wood construction. By 1874 he owned the Marquette Manufacturing Company, the largest building contractor and building supplier in Marquette and an employer of eighty men. During his long career as a mason and builder, from 1867 to the early 1900s, Gregory built all kinds of structures.

In August 1874, after the old Ridge Street School had been torn down to make way for this more substantial structure, Gregory began construction of the new high school. The school was finished in September 1875, and it cost forty thousand dollars.

The high school fronted on Ridge and Arch Streets at Pine Street. The rectangular structure with projecting pedimented central entry pavilions on both Arch and Ridge Streets stood two stories on a raised foundation. Evenly coursed hammer-dressed variegated brownstone from the quarries of the Marquette Brownstone Company rose in the exterior walls. Corner quoins, belt courses, window hoods, and sills of pure Marquette brown sandstone trimmed the school. A galvanized iron cornice encircled the building, and slate from the quarry of the Huron Bay Slate and Iron Company covered the mansard roof.[19]

Inside, a wide hall ran the length of the school building from north to south. A classroom occupied each corner of the first floor. A large high school room, two adjoining grammar school rooms (which could be converted to one room by folding the flexible doors that separated them), an equipment room, four recitation rooms, a teacher's room, principal's office, and library occupied the second floor. The basement held boilers, janitor rooms, a fuel room, and lavatories. Special attention was paid to the ventilating system, which used brick ducts attached to the chimneys that extended from the basement to the roof on each side of the hall.

As construction approached completion, the *Marquette Mining Journal* reported that the new high school would be "the finest edifice in the Upper Peninsula . . . an ornament to the city and a monument to the good sense of our people and the resources of this favored region."[20]

Saint Paul's Episcopal Church

In 1874 the Vestry of Saint Paul's Episcopal Church, Marquette's wealthiest religious community, decided to build a new stone church that would cost around thirty-three thousand dollars. The vestry appointed Peter White, W. L. Wetmore, J. C. Morse, and H. A. Downs to a building committee. The committee inspected plans, prepared estimates, and examined the prospects for raising funds. Then it reported its findings to the vestry, and the vestry adopted the plans that Gordon W. Lloyd of Detroit and Windsor, Ontario, had prepared for them the previous summer. The church would be modeled after the plan of the Episcopal church at Fond du Lac, Wisconsin, with changes and improvements as necessary, and conceived in the strictest ecclesiological terms.

Gordon W. Lloyd (1832–1905) was one of the Midwest's most fashionable church architects. He was born in Cambridge, England, and although he spent much of his youth in Canada, he trained in England under the tutelage of his uncle, Ewan Swan, a well-known church builder and restorer who practiced in the vernacular Gothic Revival and later became president of the Royal Institute of British Architects. Lloyd augmented this practical studio training with a classical education unknown in American schools by studying at night at the Royal Academy. After touring Brittany, Normandy, Touraine, and the Upper Rhine in 1856, he returned to England a refined architect committed to the northern Gothic style. Lloyd came to Detroit in 1858. There and throughout Michigan and Ohio, he soon found his unusually polished skills in great demand. Supervising construction for Lloyd was Carl F. Struck. Struck was no longer a resident partner with Gay but instead operated his own architectural practice in Marquette.

The Gothic Revival church is at 201 East Ridge Street on the site of the parish's first house of worship, a wooden structure built in 1857 and moved in 1874 to make way for the new building. Construction on Saint Paul's Church began in July 1874 and ended a year and one-half later. The church cost nearly fifty-six thousand dollars, almost twice as much as estimated. With costs so high and subscriptions difficult to attain during the panic of 1873, the building committee deferred the construction of the steps, the completion of the tower and spire, and the landscaping. Nevertheless, the church was opened for Christmas Day services in 1875.

The asymmetrical Saint Paul's Church has a steeply pitched roof, two transepts, and a buttressed and crenelated square entry tower placed at a forty-five degree angle to the southwest corner. Gothic elements appear in the stained-glass lancet windows and in the authoritative detailing in the stonework and window tracery. Executed in clear brown Marquette sandstone from the Marquette Brownstone Company quarry by the most skilled craftsmen, its rich walls of blocks of hammer-faced stone with chiseled edges, slate roof from the Huron Bay quarries at Arvon, and high-style Gothic forms celebrate both the social and economic achievements of the parish and the city.

A detailed drawing of the tower facade in the possession of Saint Paul's Church shows how the sandstone was cut and finished. The drawing specifies the exact dimensions and finishing of each block, thereby instructing the quarry on exactly what was required. The sawmill at the quarry cut and finished the stone to order. Much of the stonework for the cornice, the window, door casings, and the like was done during the winter. A man identified as Gowling, "the manipulator of fancy stone work—the boss mason of Marquette," and his force had begun placing the stone in position on the church in the spring.[21]

The church seats five hundred in two banks of butternut pews divided by a center aisle beneath an open wood truss ceiling and is illuminated by light filtered through the brilliant red, blue, and green stained-glass

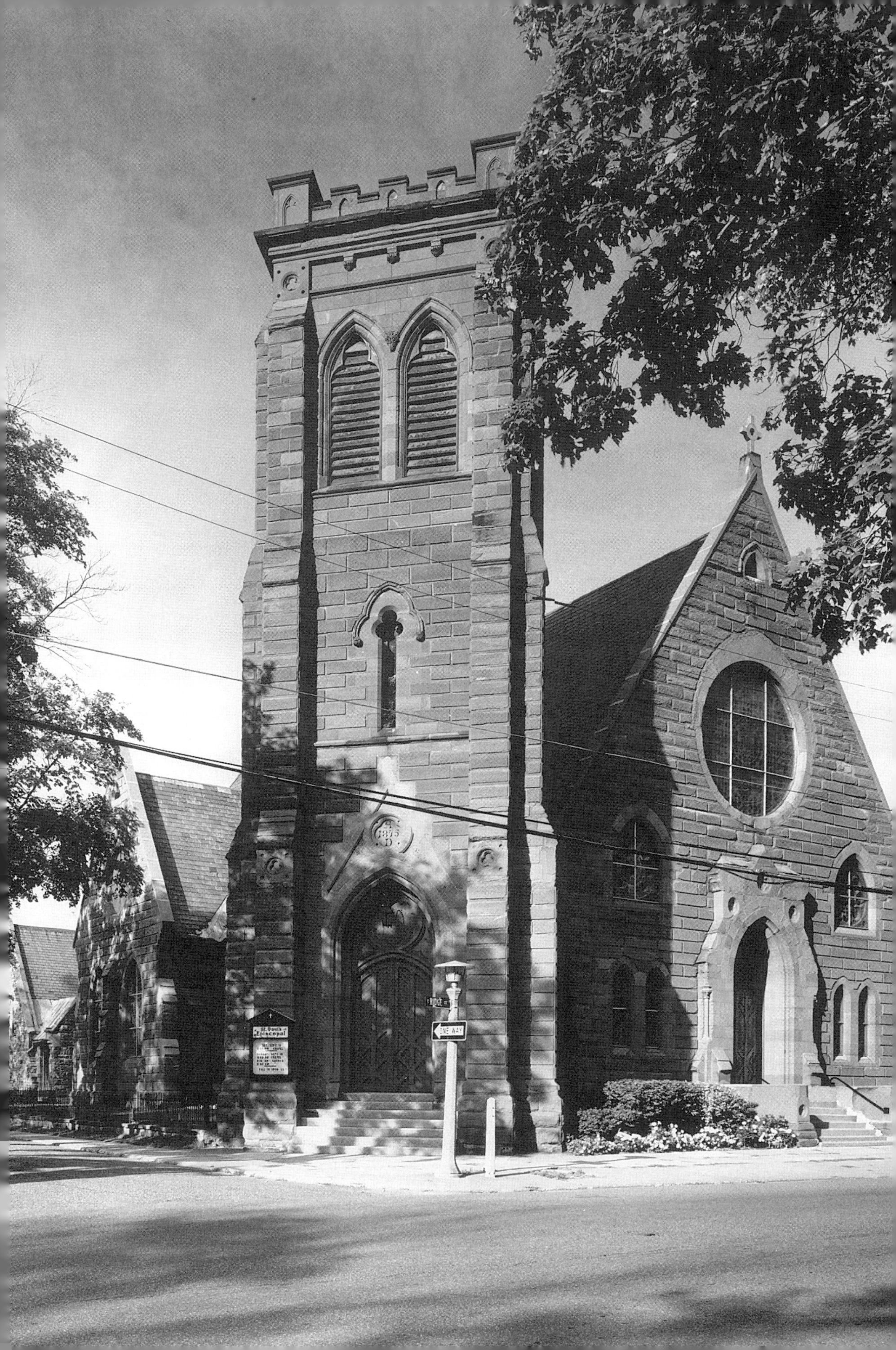
1875
St. Paul's
Episcopal
RIDGE ST
ONE WAY

windows. The woodwork was designed and crafted by A. Gustafson of Hager and Wallaster, local furniture manufacturers. Saint Paul's Church is so elegant inside and out that some regarded it as out of character with this city in the wilderness. Indeed, Saint Paul's Episcopal Church remains one of the most beautifully designed and crafted stone structures built anywhere in the Lake Superior region. The bold but fussy Postmodern addition and porte cochere testify to the continuing commitment of this congregation to its landmark church.

Morgan Memorial Chapel

Originally free standing, Saint Paul's Church is now connected by means of a hallway at the north and the corridor of the 1989–90 addition at the east to the little two-hundred-seat stone Morgan Memorial Chapel designed by the noted Chicago firm of Cobb and Frost and built in 1887–89. This chapel addition to Saint Paul's Church was commissioned by Peter and Ellen White in memory of their eldest son, Morgan Lewis White, who died in 1875 at age twelve. It is a single-gabled chapel with a south entrance porch. Buttresses support its rock-faced variegated Marquette sandstone walls. *The Resurrection Window*, a stained-glass creation manufactured by Tiffany, fills the large Gothic opening in the front (west) gable. Its deep reds and blues fade in the center to pale greens, blues, and browns, and the whole contrasts with the rich wood of the hammer-beam ceiling and of the paneling.

Saint Paul's Episcopal Church and the Morgan Chapel invite comparison with the reddish brown Lake Superior sandstone Saint Paul's Episcopal Church in Milwaukee. The Richardsonian Romanesque Milwaukee church used randomly coursed rock-faced red Basswood Island sandstone. The Milwaukee church is only one example in which red Lake Superior sandstone was employed as a building material in the cities of the Upper Great Lakes outside the Jacobsville formation and the Bayfield group and elsewhere in America.

Early in 1882, the building committee of the vestry of Saint Paul's Episcopal Church, Milwaukee's oldest Episcopal parish, which had organized in 1836, invited five prominent architectural firms to submit designs for a new church at 904 East Knapp Street (the northeast corner of North Marshall Street). The new church would replace the parish's pioneer wooden Gothic Revival church, built by George Mygatt in 1844–45. On 13 May 1882 the committee selected the plans for an impressive red sand-

Facing page: Saint Paul's Episcopal Church, 1874–77, 1887, Gordon W. Lloyd, Marquette, photograph c. 1985. (Courtesy Balthazar Korab)

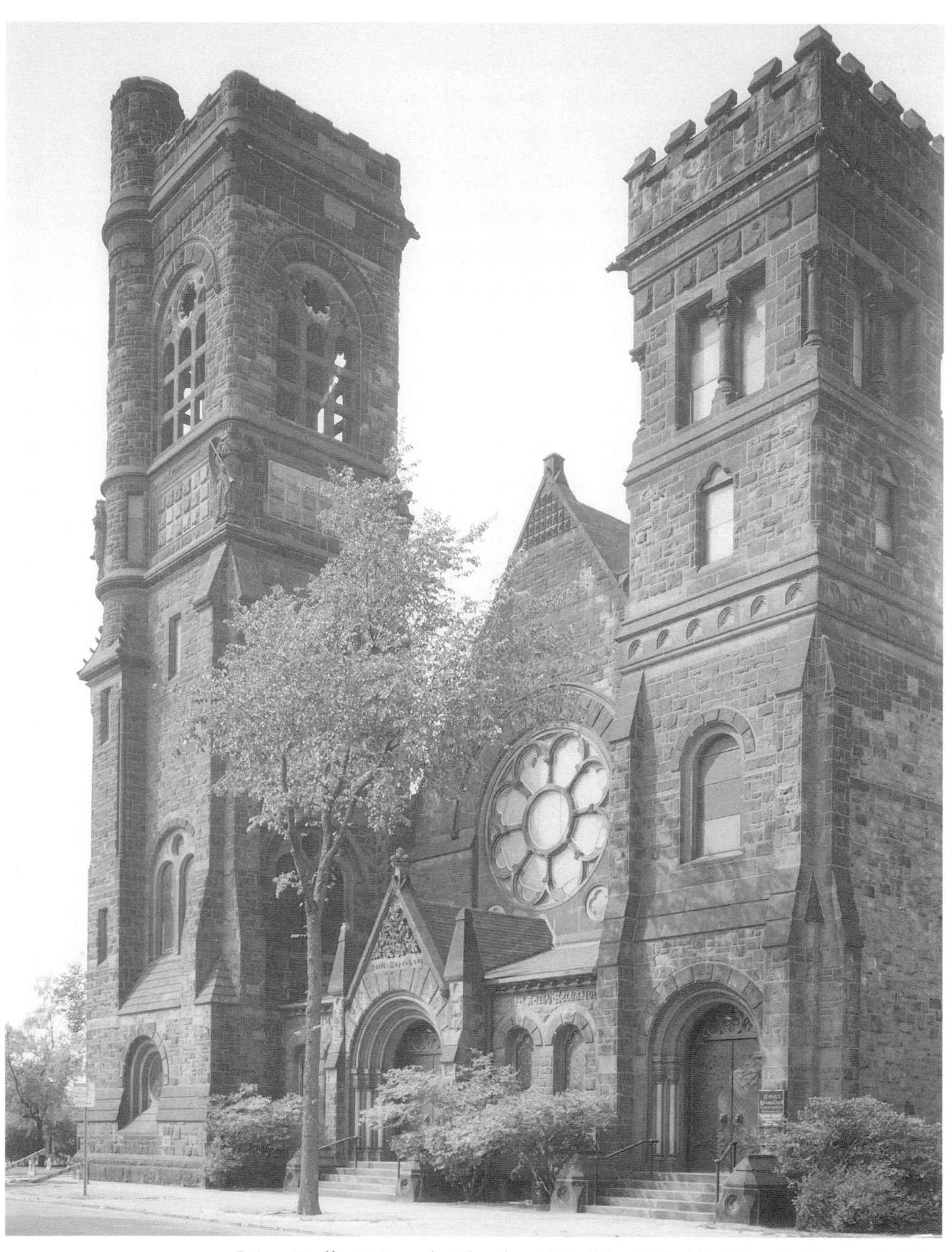

Saint Paul's Episcopal Church, 1882–85, 1888–89, Edward Townsend Mix, Milwaukee. (Courtesy Library of Congress, Prints and Photographs Division, HABS WIS, 40-MILWA,35-1, Jack E. Boucher, 1969)

stone Richardsonian Romanesque church prepared by Edward Townsend Mix (1831–90).[22] Mix's competitors for this church design included Ware and Van Brunt of Boston, Howland Russell of Milwaukee, and Richard Upjohn and Henry G. Harrison of New York City. On learning he had won the commission, Mix set out with Edwin Hyde, Milwaukee's eminent stone merchant and contractor, for Basswood Island to select the stone for the church themselves.

Mix based his design for Saint Paul's Church on Henry Hobson Richardson's unexecuted design for Trinity Church, Buffalo, which was published in the *Architectural Sketch Book* for July 1873. The dimensions of the church are 140 feet by 84 feet. At the southwest corner, a tower rises to lofty heights. A carved stone angel with trumpet, a motif derived from Richardson's Brattle Square Church in Boston, decorates each of the tower's corners. The walls are rock-faced, deep red Basswood Island sand stone. Mullions and tracery are yellow sandstone; engaged colonnettes flanking three main entrances and columns on the west porch are polished granite. Richly colored Tiffany glass windows adorn the interior.

The sandstone was shipped from Basswood Island by Hibbard and Vance, cut by Cook and Hyde, and laid in the walls by John G. Jones. F. A. Purdy, under the supervision of Mix and a Mr. Loeher of New York, carved the four stone angels ornamenting the tower.

Excavations for the foundation of Saint Paul's began in the fall of 1882. The cornerstone was laid on 17 June 1884, and services were first held in the still-unfinished church on 12 October 1884. In 1884–85 the chapel was completed, in 1888–89 the southwest corner tower completed, and in 1890 the pale yellow Milwaukee brick parish house completed. The Episcopalian community consecrated the church on 11 November 1891. The project cost $230,000.[23] This splendid church for a pioneer parish built of rugged Basswood Island sandstone to the designs of a popular local architect speaks of Milwaukee's identity and pride. The necessity of obtaining first-class reddish brown Lake Superior sandstone for its exterior surfaces brought Cook and Hyde to Basswood Island.

Both Saint Paul's Episcopal Church in Marquette and Saint Paul's Episcopal Church in Milwaukee employed the highest grade of sandstone available, the former rich brown hammer-dressed sandstone from Marquette, the latter, rock-faced rosy red sandstone from the Apostle Islands.

Saint Peter's Roman Catholic Cathedral

The original Saint Peter's Roman Catholic Cathedral, built in 1864 at 311 West Baraga Avenue and itself Gothic, was destroyed by fire in 1879. The Diocese of Marquette replaced the burned-out structure with the present

Saint Peter's Roman Catholic Cathedral and Baraga Central High School, 1903, John D. Chubb, Marquette, photograph c. 1905. (Courtesy Library of Congress, Detroit Publishing Company Photographic Collection)

twin-towered Romanesque Revival church. Constructed of variegated Marquette brownstone from the South Marquette quarry, it took ten years to complete at a cost of one hundred thousand dollars. Since no architect capable of creating a structure of this magnitude and grandeur resided in the Upper Peninsula in the 1880s, the church was designed by Henry G. Koch and Son, a noted Milwaukee firm. Although a Wisconsin city, Milwaukee was close to the Upper Peninsula, less than one day away by train on the Chicago and Northwestern Railroad, and had as much if not more of a commercial affinity with the cities of that region than Detroit.

The church measures one hundred feet in length by eighty feet in width, and triple arches mark the main entrance of the building. Inside there is a single-aisled, barrel-vaulted nave and two small side aisles and short transepts. The church was profusely decorated with memorial cathedral

Facing page: Saint Peter's Roman Catholic Cathedral, 1880, Henry G. Koch and Son, 1935 rebuilding, Edward A. Schilling, Marquette, photograph c. 1985. (Courtesy Balthazar Korab)

glass windows from George A. Misch of Chicago and with woodwork from Hager and Johnson of Marquette. The cost and beauty of its material, its large size, and the elegance of its design and finish reflect the struggle of the Catholic church, through the Diocese of Marquette, to establish its identity in a burgeoning youthful city on the American frontier. In 1935, after another fire, the church was rebuilt extensively to the plans of Detroit architect Edward A. Schilling.[24]

Using Marquette brownstone salvaged from the Grace Furnace—erected by the Lake Superior Iron Company of Ishpeming in 1871–72 in Marquette's Lower Harbor and demolished by the Lake Shore Engine works in 1902—the diocese built Bishop Baraga Central High School to the plans of John D. Chubb of Chicago in 1903. The school was destroyed in 1975.

Daniel H. and Harriet Afford Merritt House

The economic boom of the 1880s sparked a surge of building activity. Marquette's population reached ten thousand in 1890, and the many fine residences and commercial buildings attested to the city's prosperity. In 1880, during the economic boom, Daniel H. Merritt, a pioneer industrialist and an employee of the Lake Shore Iron Works (owner of the Iron Bay foundry and machine shops), commissioned Hampson Gregory to build for himself and his family an elegant brownstone Italianate house on the south side of Ridge Street where it would command a view of Lake Superior, the city, and the surrounding country. The Daniel H. and Harriet Afford Merritt House stands at 410 Ridge Street.

Born in Green County, New York, Merritt (1833–1919) came to Marquette in 1858 to work for the Iron Mountain Railroad. He became the general manager of its successor, the Marquette, Houghton and Ontonagon Railroad, a position he held until 1874. From 1868 to 1893, Merritt also worked for the Lake Shore Iron Works. Merritt invested in mining companies, banks, and land. When the Merritts built their fifteen-room house, they belonged to Saint Paul's Episcopal Church and held a prominent position in social circles.

The house was designed by English immigrant Hampson Gregory. Steeped in the stone building tradition of his native Devonshire, Gregory built it of roughly dressed, evenly coursed, variegated reddish brown and white sandstone. By this time, Gregory had designed and built many stone buildings in Marquette and had promoted himself from builder to architect and superintendent. At the middle of the decade, the newspaper would note that Gregory, more than any other man, had been responsible for the building of Marquette.[25] Gregory was the very best choice for a client seeking the services of a local man capable not only of designing and building but handling stone.

Daniel H. and Harriet Afford Merritt House, 1880, Hampson Gregory, Marquette, photograph c. 1985. (Courtesy Balthazar Korab)

Hiram A. and Sarah Benedict Burt House, 1872–76, Carl F. Struck, Marquette, photograph date unknown. (Courtesy Jack Deo, Superior View)

As construction began, the Marquette newspaper predicted the Merritt house would enhance the appearance of an already beautiful residential street and would not only be the handsomest but the costliest private residence in the Upper Peninsula.[26]

A center tower topped by a concave mansard roof and displaying a wrought-iron balconet covered by a bracketed hood projects from the nearly square Italianate house with Second Empire detailing. A porch supported by simple columns and ornamental scrolled brackets wraps around the east and south sides of the house at the ground floor.

The ground floor contained the reception room, billiard room, laundry, pantry, kitchen, and bathrooms. The second floor held the library, parlor, dining room, bath, and bedroom. The third floor had ten sleeping apartments and a bath. The interior is fitted with walnut and cherry woodwork finished in Marquette shops. Today the house is divided into apartments, but the exterior and many interior features remain unchanged.

A few doors away at the southeast corner of Ridge and Blaker Streets, Gregory had constructed the Italian Villa with Second Empire detailing for Hiram and Sarah Benedict Burt eight years earlier, in 1872–76. This house is unique among his many Marquette projects. The exterior walls are skillfully laid with hexagonal blocks of sandstone, presumably from the Burt Freestone Company quarry in South Marquette. From this location Hiram Burt, prominent in mining and iron interests, overlooked the harbor for which he had promoted improvements.

Harlow Block

During the intense building activity of the 1880s, Amos R. Harlow (1815–90), a founder of Marquette, also commissioned the local stonemason and builder, Hampson Gregory, to create the Harlow Block, a commercial building at the northwest corner of Washington and Front Streets, the principal intersection of Marquette's downtown commercial district.

Harlow was admired for his real estate developments that built up and improved the city and afforded employment.[27] He built the Harlow Block as speculative office and store space. It gave tenants a good address in a substantial building located in the very heart of the downtown business district.

The Harlow Block is a straightforward commercial design of the Victorian era. Constructed in 1887 of plain, smooth-sawn, variegated Marquette brownstone locally called "raindrop" for its purplish brown iridescence, it is a rather heavy-handed version of the Italianate style that marked Main Street America during the 1860s and 1870s.

The first floor contains five stores, three facing Washington Street, two facing Front Street. All display large plate glass windows and iron fronts. Entries from each street lead to intersecting hallways. The second and third floors contain offices and rooms. All are supplied with water, ventilating flues, light from at least two windows, and steam heat. Solid stone walls divide the basement into store cellars. From one entrance on the alleyway, goods arriving for the stores could be sent to the cellars without disturbing customers. The building was built fireproof: solid masonry fire walls rose from the basement to the roof between stores, iron doors separated rooms from hallways, vaults stood on every floor, fire escapes led from every floor, and solid masonry formed walls and partitions.

The completion of substantial stone buildings like the Harlow Block led one writer to conclude that Marquette gives the impression of "a large city in miniature." Unlike most rapidly growing western towns that resembled "an uncouth boy, whose clothing has unfortunately far from kept pace with his growth," Marquette seemed by 1887 to combine the vigor and roughness of the West with the polish and sophistication of the East. This, the writer explained, was due in large part to paved streets, progressive business activity, costly and beautiful church architecture, substantial and convenient schools, and the handsome business blocks of brick or locally quarried freestone.[28]

Upper Peninsula Branch Prison and House of Correction (Marquette Prison)

Established by the Michigan legislature in 1885, the Upper Peninsula Branch Prison and House of Correction at Marquette stands on a site in Marquette thought "too beautiful for a prison."[29] The site was selected over sites at Escanaba, Negaunee, and L'Anse, however, because of the city's location, advanced development, and aggressive businessmen.[30] The erection of the Marquette prison in 1885–89 marked Marquette's important role as a governmental center for the Upper Peninsula.

The Michigan legislature established the Marquette prison because the increased criminal population of the state overcrowded the Michigan Asylum for Insane Criminals at Ionia and the Detroit House of Correction and because county jails were no longer regarded as effective accommodations in which to rehabilitate criminals. Act 148 of the (Michigan) Public Acts of 1885 established the Upper Peninsula Branch State Prison and House of Corrections and appropriated funds for its location and erection. Governor Russell A. Alger appointed a blue ribbon board of prominent Upper Peninsula citizens representing all geographic areas and many industries and businesses to investigate and to manage the construction of the prison. The six-member Board of Commissioners included Peter A. Van Bergen of Menominee, Eli P. Royce of Escanaba, James M. Wilkinson of Marquette, Eli B. Chamberlain of Saint Ignace, Charles Hebard of Pequaming, and John Duncan of Calumet. The board selected and secured a site; adopted plans and specifications for the grounds, buildings, and fixtures; and reviewed bids and proposals for construction. The intent was to design a facility in which prisoners could be punished with strict military discipline and corrected by productive work.

At the very outset, a Marquette quarryman urged the board to select an architect whose design for the prison would call for Portage Entry red sandstone. This happened only a few months after the legislature had appropriated funds for the prison and after the board had elected officers and published in the newspaper a request for the donation of a site. John H. Jacobs, the owner and manager of the Wolf and Jacobs quarries recently opened at Portage Entry, asked James M. Wilkinson, treasurer of the board, to give full consideration to John Scott of William Scott and Company, architects:

> Mr. John Scott of the firm of Wm Scott & co Architects of this city [Detroit] will call on you and make an application for making the Plans

> of the new States prison that will be built at Marquette I trust he is the best friend our Portage Entry Red Stone has got and about the only friend so what helps me will help you the Portage is what I want to figure on when the job is to be let and if you can help him to get the making of the plans it will be of grate [*sic*] help to us the Royalty is quite a consideration if possabely [*sic*] you will do all you can for him his Partner has bin [*sic*] to see the Portage Quarry last year and likes the stone Well the sale of the stone depends on the success of the quarry.
>
> They are the oldest and most reliable firm that I know off [*sic*] and have done a grate [*sic*] deal of this kind of work they made plans of the Pontiac jail and used our stone. I trust you will do for Mr. Scott all you can and oblige.[31]

The board visited the penal institutions at Jackson, Ionia, and Detroit and met with the State Board of Corrections and Charities and with prison wardens to learn about prison architecture. The board chose the designs and plans made by William Scott and Company. The plans were modeled largely after the Ionia House of Correction.[32] The board employed Scott and Company to supervise construction.

The Scott firm was gaining experience in the design of public buildings as state and local governments were building their first substantial schools, insane asylums, orphanages, courthouses, jails, and city halls in recently settled frontier areas. It won several important public and private commissions in the Upper Peninsula in the late 1880s and early 1890s and used the indigenous red and brown Jacobsville sandstone in most of them. In fact, John Scott's first concepts for the Wayne County Courthouse (1896–1902) in Detroit would call for Jacobsville sandstone.

When William Scott's plans for the prison were selected in 1887–88, the firm opened a branch office in Marquette and placed Demetrius Frederick Charlton in charge of the office and the prison construction. William Scott, a trained architect, had emigrated from Ipswich, England, to Windsor, Ontario, in 1853. In 1875 he located in Detroit and formed a partnership with his two sons, John and Arthur Scott. After William Scott retired in 1889, the firm changed its name to John Scott and Company, but in Marquette it was known as Scott and Charlton.

Once the prison was completed, Charlton remained in Marquette, quickly becoming the most prominent architect in the Upper Peninsula. Charlton (1856–1941) was born in Ingtham, Kent County, England. He immigrated first to Canada and then to Detroit in 1877. By 1881 he was working as a draftsman for Gordon W. Lloyd, and by 1885, for William E. Brown. In 1887 he married Cornish-born Alice H. Grylls, the sister of H. J. Maxwell Grylls, who also was employed by John Scott but later became a partner in Smith, Hinchman, and Grylls. Trained as a civil engineer,

Upper Peninsula Branch Prison and House of Correction (Marquette Prison) 1885–89, William Scott and Company with Demetrius Frederick Charlton, Marquette, photograph c. 1890. (Courtesy Marquette County Historical Society)

Charlton was elected a Fellow of the American Institute of Architects in the spring of 1893. He worked for the passage of the first state registration act and was appointed by the governor to its first board.[33]

In 1892 in Marquette, Charlton took in an associate, R. William Gilbert, and in 1895 another, Edward Demar. The firm opened branch offices in Sault Sainte Marie and Houghton, Michigan, and in West Superior, Wisconsin. By 1903 Edwin O. Kuenzli joined Charlton, and they had a branch office in Milwaukee. This association continued until Charlton retired in 1917. By then, Charlton had designed eighty schools, twenty public buildings, three Carnegie libraries, and two hundred fifty business blocks and houses.[34]

The Marquette Prison was designed, constructed, and furnished at a cost of $250,000. The original appropriation of $150,000 for the prison contemplated the completion of only one cell wing and did not provide heat, light, or furnishings. In 1888 the Michigan legislature appropriated $75,512 for completing and furnishing the building and for constructing the second cell block.[35]

The High Victorian Romanesque administration structure with square central tower and octagonal turrets is constructed of reddish brown Marquette sandstone trimmed with red Portage Entry sandstone and covered with an immense hipped roof of Lake Superior slate. The rotunda and the cell block wings were built of Marquette brownstone backed by stone quarried on the prison site.

The *Marquette Mining Journal* for 28 April 1888 described in great detail the prison complex while under construction:

> In an architectural sense, the administration building is the building of all. . . . It is to be built entirely of Marquette and Portage Entry rock-faced sandstone, and will present the most elegant and massive appearance of any building north of the straits. . . . Marquette brownstone will be used entirely except for the string course, cornices, window arches, gables, the embattlements of the tower, etc., which will be of Portage Entry red sandstone. The roof will be of Lake Superior slate, which was used on all the other buildings, with terra cotta ridging. On the building there will be a great deal of carved and dressed stone work; this is mostly finished and ready for the builders, the stone cutters having been at work all winter. Some of it is very beautiful and required an immense amount of work.

Only the administration building, northwest cell block wing, and rotunda remain of the original walled complex, but the lovely historic gardens before the administration building have been restored.

The administration building contained the residence for the warden and his family as well as his business office and offices and living quarters for the deputy warden, physician, chaplain, and other officers. From the administration building, a long corridor led to the rotunda. The first floor of the rotunda contained a guard room, which permitted views of both cell block wings and offices. Two broad stairways led to the vast hall of the second floor, unobstructed by the truss-supported roof. A small gallery over the stairs provided space for a chapel.

At Marquette the prisoner would work during the day in a common area making brooms, snowshoes, or cigars—or he would improve and beautify the prison grounds—and he would sleep at night in a single cell. Marquette prison follows the Auburn, named for the prison at Auburn, New York (1816–25), rather than the eastern penal system, named for the Eastern Penitentiary at Cherry Hill near Philadelphia (1825). Auburn advocated the rehabilitation of criminals through a regime of nighttime sleep in cells and daytime work in common areas. Eastern kept prisoners in solitary confinement day and night.

Behind the prison stood the dining and hospital building. Nearby was the engine house that furnished heat and light for the prison. An arched conduit of brick ran through all of the buildings carrying the water and steam pipes, speaking tubes, and electrical wires.

The prison's fortress-like medieval style, its substantial sandstone walls, and its efficient plan together expressed current thought on the punishment and reform of criminals.

John Munro and Mary Beecher Longyear House

The largest and most carefully planned house of Lake Superior sandstone ever constructed in the region was the John Munro and Mary Beecher Longyear House. From 1892 to 1903, the house occupied a one-block site at the corner of Arch and Cedar Streets on the high rocky bluff overlooking the rough waters of Lake Superior. Although open to views of the lake from every direction, the house was "presentable" on the west to the city. A stone retaining wall surrounded the property. The architect was Demetrius Frederick Charlton, and the landscape architect was the renowned Frederick Law Olmsted.

James Monro Longyear (1850–1922) came to the Upper Peninsula from Lansing in 1873 to explore and examine mineral and timber lands for the state government, investors, and land, lumber, and mining companies. He initially secured surveying jobs through the influence of his father, who was a lawyer, congressmen, and judge. Later, in 1879, young Longyear's extensive knowledge of the wild lands of the Upper Peninsula led to his appointment as land agent for the Keweenaw Canal Company, builder of the canal that connected Portage Lake with Lake Superior. He settled in Marquette. Longyear's shrewd ability to assess the region's mineral and timber resources and to deal with speculators, developers, and regulators aided his own transactions. Investments in the Gogebic, Menominee, and Mesabi Iron Ranges, dealings with the Steel Trust, and a coal-mining venture in Spitzbergen, Norway, made him rich and brought him into contact with some of America's foremost capitalists and industrialists. By 1890, when the Longyears built their house, their family had grown to five children and a household of servants, and their fortune to millions. Travel throughout the United States and Europe had expanded Longyear's vision. He then served as mayor of Marquette.[36]

On 25 January 1890 the *Marquette Mining Journal* announced the Longyears' intent to erect a large expensive house on a lot adjoining their present wooden Queen Anne house at 425 Cedar Street. That winter and the following spring the Longyears developed the plans and design for their house and chose an architect to carry them out. They invited Eugene C. Gardner of Springfield, Massachusetts, Treat and Foltz of Chicago, and Charlton to submit plans and sketches.[37] Mr. Longyear described the kind of house they wanted in a letter to Gardner of 14 February 1890: "My idea is that the house will cost say $40,000. We want a 'first-class job' if we build, and the above named figure is a guess at the cost. . . . My idea is to build a large country house."[38]

By means of letters, memoranda, and diagrams, Longyear conveyed his instructions and those of Mrs. Longyear to Gardner and to Treat and Foltz but did not ask them to visit the site.[39] Longyear detailed their pref-

John Munro and Mary Beecher Longyear House, 1890–92, Demetrius Frederick Charlton, Marquette, photograph 1892. 1903–4, dismantled, transported, and reassembled in Brookline, Massachusetts. View from Cedar Street and the city. (Courtesy Marquette County Historical Society)

erence for the placement of the house on the site, the arrangement of interior space to accommodate their family, servants, and guests, and the appearance of the exterior. The absence of written correspondence between Longyear and Charlton suggests that Longyear spoke directly with Charlton since they lived in the same city.

The Longyears paid special attention to the impression the physical appearance of the exterior would have on the viewer. After all three firms had prepared satisfactory floor plans, Longyear asked each to submit accompanying elevation sketches. In a letter to Gardner of 19 March 1890, Longyear explained, "there are fine views of the house in all directions, except the west, but it is important that the west side should be

Longyear house, photograph 1892. View from Lake Superior and Ridge Street. At top right is the previous Longyear house. (Courtesy Marquette County Historical Society)

presentable as the town lies in that direction and the house is approached from that way."[40] In the same letter, Longyear called for four fronts since the house would be entirely isolated from other buildings and stand in a conspicuous position on the point of the ridge: the south side would face the harbor, the east would look out over Lake Superior, the north would front the street from which all carriages must approach, and the west would be visible from the entire length of Cedar Street, the neighborhood's most traveled road. Longyear stated that they preferred no one style to another. They did not like "gingerbread," however, but wanted the exterior beauty to consist mainly of proportions, lines, and, we assume, texture and color.

The Longyears selected the plans drafted by Demetrius Frederick Charlton, which offered a choice between a Scotch Baronial and a Romanesque exterior without affecting the interior arrangement, and rejected the plans of the other two firms.

Construction on the house began around 14 June 1890. Contracts were awarded most frequently to local people, even when they were not the low bidder. John H. Jacobs won the contract for supplying the brownstone for the house from Marquette quarries. Charles Van Iderstine laid the

Longyear family, architect, and construction crew on porch of the Longyear house, photograph 1892. From left is Charlton, Longyear, Mrs. Longyear, Longyear child, and Van Iderstine. (Courtesy Marquette County Historical Society)

stone and superintended the construction of the house.[41] The house was completed on 8 October 1892.

The house was constructed entirely of Upper Peninsula materials—Lake Superior red and brownstone, Huron Bay slate, and native woods, all handled with robust directness. An octagonal tower on the north corner, projecting pavilions, decorative chimneys, and a porte cochere with attached two-story tower pushed out from the solid rectangular mass. A large veranda swept around the east and south sides of the house. Short columns supported the semicircular entrance arch that faced northwest from Cedar Street.

Richly detailed and finished, the interior was arranged around a great octagonal hall open to the roof. Around the great hall pivoted the dining room, entrance hall, reception room, and library. The octagonal-ended dining room with many windows, a fireplace, and inglenook stood on the

Longyear house. Citizens saw the reputation of Marquette derived from its fine collection of sandstone architecture. From Marquette, Michigan. Illustrated *(Milwaukee: Cramer, Aikens and Cramer, 1891), 24.*

northeast corner. The rooms most frequently used by the family—the sitting room, breakfast room, and library—overlooked Lake Superior. On the second floor, a gallery surrounded the great octagonal hall. This floor and the attic held bedrooms, bathrooms, and servants' quarters. The basement contained the kitchen, laundries, and boiler room. Many open fireplaces supplemented the steam heat and made the house comfortable. The detached tower contained the children's playroom on the first floor and a photography room and laboratory on the second.

In its bold geological forms and in its sensitive response to the nature of materials, the Longyear house was comparable to the contemporary James J. Hill Mansion in Saint Paul, Minnesota. Designed by Peabody, Stearns and Furler of Boston, this huge house is one of the most dynamic examples of Richardsonian masonry in the upper Middle West. The principal difference between the two is that the Longyear house was more modest, reflecting the less expansive economic situation in Marquette.

At the suggestion of his friend, Theodore M. Davis of Newport, Rhode Island, who was president of the Keweenaw Canal Company, Longyear sought the advice of Richard Codman of Boston with regard to the

Longyear house rebuilt at Brookline, Massachusetts. From the Longyear scrapbook. (1909). (Courtesy Longyear Museum)

interior finish and decoration. From Codman the Longyears purchased lighting and fireplace fixtures, wallpaper, Bulgarian embroideries, Vantine rugs, Cottier furniture, mirrors, and clocks.[42]

Frederick Law Olmsted (1822–1903) of Brookline, Massachusetts, America's greatest landscape architect at the time, inspected the site and planned the landscaping. Into the scheme he incorporated stone retaining walls, iron fences, enclosed courts, stone sidewalks and stairways, tennis courts, a gymnasium, and an arbor and native plantings.[43] As the landscaping plans were revealed to the public on 23 April 1893, the *Marquette Mining Journal* declared, "The Longyear residence . . . will stand as the finest house in all Michigan and the peer of any palatial residence in the Northwest." Indeed, one biography of Longyear refers to his magnificent stone residence on the shore of Lake Superior, "one of the most beautiful and costly homes in the state of Michigan," as evidence of the stature of the man.[44] Handbooks and citizens' guides to the city of Marquette referred to Longyear's palatial sandstone house as evidence of the stature of the city.

The huge scale in the Richardsonian Romanesque style; magnificent view of Presque Isle, Lake Superior, and the city of Marquette; native sandstone and wood building materials; and superb workmanship made the Longyear house the equal of any in the upper Middle West.

After their son drowned in Lake Superior and after a trip to Paris in 1902, the Longyears decided to move east. In 1903 the house was dismantled, loaded onto railroad cars, and transported east to Brookline, Massachusetts, for rebuilding. Charlton and Van Iderstine oversaw the move and rebuilding. Ripley's "Believe It Or Not" wrote up the moving process. In 1937 the house was opened to the public as the Longyear Museum, later the Mary Baker Eddy Museum. Today the house is the centerpiece of a gated condominium complex called Longyear at Fisher Hill.[45]

Marquette County Savings Bank

In 1890, as Marquette was emerging from a major population and construction boom, Nathan M. Kaufman (1862–1918), a merchant and speculator in mines, and other leading Marquette businessmen organized the Marquette County Savings Bank as a convenience for small depositors and to facilitate real estate loans. The two banks then in existence served the extensive commercial demands of the city; the savings bank would serve smaller depositors and borrowers—the working people who constituted the majority of the population.

The bank's incorporators included some of the hardest working and foremost businessmen of the period in Marquette: C. H. Call, a businessman and the bank's first president; William F. Fitch, general manager of the Duluth, South Shore and Atlantic Railroad Company at Marquette; Samuel R. Kaufman, one of the oldest and most successful merchants in the city; W. P. Healy, an attorney and one-time member of the Michigan legislature; Mary Breitung, widow of Edward Breitung, a major entrepreneur in the Upper Peninsula mining industry; and Nathan M. Kaufman, general manager of the Breitung estate, then one of the largest holders of mineral lands in the Upper Peninsula.[46] M. R. O'Brien, president of the People's State Bank in Detroit, and Samuel Mitchell of Negaunee, were also incorporators. Except for Samuel R. Kaufman, the incorporators became the first directors. Of particular influence in the execution of the bank building project was Nathan M. Kaufman. Born to pioneer settlers Samuel R. and Juliette Graveraet Kaufman, the young Kaufman carried on the family dry-goods business before investing in iron mines and entering the financial industry.[47]

Not until all stock, amounting to fifty thousand dollars, was subscribed and paid for, and not until the company was organized fully under the laws of the state, did the promoters of the bank reveal their plans. The *Marquette Mining Journal* announced the organization of the savings bank

Nathan M. Kaufman (1862–1918). From Memorial Record of the Northern Peninsula of Michigan *(Chicago: Lewis Publishing Co., 1895), 112.*

on 5 July 1890, noting that "it is probable that no enterprise of equal magnitude was ever before organized so quietly and quickly in Marquette." With such sound financial backing and strong management, the new bank would enjoy the confidence and business of the public.

By August 1890 the Marquette County Savings Bank opened in the Manhard Block on Front Street. Within four and one-half months, the bank's deposits rose to $196,000, and the directors made plans to abandon these quarters and construct a modern substantial fireproof bank and office building. In January 1891 they purchased for $14,000 the site at the southeast corner of Washington and Front Streets known as Peck's corner and then occupied by a block that they planned to demolish.[48]

Barber and Barber, a Duluth firm with no previous connections in Upper Michigan designed the bank, and Noble and Benson of Merriam Park, Saint Paul, built it.[49] Hager and Johnson Manufacturing Company of Marquette furnished the interior finish of quarter-sawn oak. Including the land, construction, and finishings, the bank represented an investment of one hundred thousand dollars.

Located at the principal intersection of downtown Marquette on the southeast corner of Front and Washington Streets, the bank building is on a site sloping east toward the lake and south toward the railroad's final approach to the ore docks. The steel-frame, fireproof structure rises five stories at Front Street, six at Washington Street, and seven at the rear. The outer walls of the first floor are Rock River brownstone laid in rock-faced ashlar. Above that they are red pressed brick. Semicircular bays, the northwest of which terminates in a tower containing a three-faced illuminated Howard clock, rise from the second story to the flat roof and flank the pedimented central entry section. This is decorated by stone spandrel panels carved with stylized foliated designs. The recessed entry is supported by polished granite columns with carved capitals. At the time, according to the *Marquette Mining Journal* for 18 April 1891, the building was termed "modern American with Gothic feeling," but is, in fact, a fascinating commercialized version of the Queen Anne style.

The elevator and ventilating shafts, staircase, and lavatories are arranged on the south wall of the bank building to deaden the noise of the ore cars that once passed through the Jackson cut to the docks. A central hall runs the length of the structure on each floor, giving access to offices on the north and west of the upper floors. The banking rooms occupied the main floor. The basement contained safety deposit vaults, offices, and a barber shop. A subbasement held the janitor's quarters in the east portion, which opens directly outside, and boiler and engine rooms and coal storage in the west portion, which is entirely below ground.

Special attention was paid to making the building fireproof. The building stood alone and isolated from others. Construction with iron columns, steel beams, and fireproof tiling for the floor arches and partitions extended through all the floors, the upper ceiling, and the roof. The only combustible materials were the oak window and door casings and one-inch-thick maple flooring laid on cement.

In February 1892, when the Marquette County Savings Bank opened the doors of its "elegant fire proof building" on the main corner of Marquette, it was acknowledged to be "the first thoroughly modern and metropolitan fireproof business and office block ever built in this part of Michigan."[50] The wild and flamboyant modern bank building symbolized the soundness and permanence of this financial institution and the role the working people played in the growth and prosperity of Marquette.

Although scaled to the economy of Marquette, the Marquette County Savings Bank building resembles the many red Lake Superior sandstone office and bank structures built in the larger cities of the Upper Great Lakes

Northwestern Guaranty Loan (Metropolitan) Building, 1890, Edward Townsend Mix, Minneapolis, photograph date unknown. Destroyed 1961–62. (Courtesy Minnesota Historical Society)

and elsewhere in America. For example, in 1890, two years after establishing the Northwestern Guaranty Loan Company, Louis F. Menage, a real estate speculator, built a large Richardsonian Romanesque structure to house the company and other office tenants in Minneapolis. The building was located at the southwest corner of Third Street and Second Avenue. Edward Townsend Mix of Milwaukee was the architect. The Minneapolis office building was constructed almost entirely of sandstone rather than of brick trimmed with sandstone like the Marquette office and bank building. The exterior walls of rock-faced red Lake Superior sandstone rose in all four sides of the Northwestern Guaranty Loan (Metropolitan Building) above a base of green New Hampshire granite. Like the Marquette County Savings Bank, a corner tower soared forty feet to a belvedere, and carving embellished the main entrances. The interior of the Minneapolis structure was arranged around a magnificent glass and iron light court from the second floor to the roof. Unfortunately, the Northwestern Guaranty Loan (Metropolitan) Building was lost to urban renewal in 1961–62, while the Marquette County Savings Bank building has been beautifully restored for use as professional offices by its current owners, Peter Kelly and Robert Berube, under the provisions of the historic preservation tax credits.

Marquette City Hall

By 1893 Marquette had an excellent school system, a public library, electric lights, public transportation, fire protection, a public water system, and a large public park at Presque Isle. But it had no city hall.

That year, the people of Marquette began plans to build a city hall with local labor and local materials. This monumental government building is a blend of Second Empire, Renaissance Revival, and Richardsonian Romanesque styles. It was constructed by Emil Bruce in red brick and purplish brown sandstone after designs by Andrew W. Lovejoy and Edward Demar of Marquette.

During the panic of 1893, newly elected Mayor Nathan M. Kaufman convinced the city council and citizens to issue bonds for fifty thousand dollars to build a city hall with local labor and local materials, thereby putting some Marquette men back to work. In the spring of 1893, when Peter White notified the Common Council of the City of Marquette that the city's rent-free lease on his building on Spring Street it then occupied as a city hall had expired, and the city would need to vacate the building, the council decided the city should build a suitable building for the city and its departments. The council instructed the Committee on Parks, Cemeteries, Public Buildings, and Grounds to obtain plans and estimates for a spacious and conveniently arranged city hall.[51]

Within one month, on 25 May 1893, Kaufman called a special council meeting to consider building a city hall and bonding the city to raise the necessary funds. The council resolved to build a city hall, not so much because the city needed to find other quarters in which to hold meetings and safely keep records, but because, it explained, smaller less prosperous and important cities than Marquette owned a handsome city government building, and because both public economy and public necessity required the construction of a city hall. The government build-

United States Custom House, Post Office, Court House, and Government Building, 1888–89, M. E. Bell and Will Freret, and Marquette City Hall, Marquette, photograph 1905. The Federal Building is on the right. (Courtesy Library of Congress, Detroit Publishing Company Photographic Collection)

ing the council envisioned would be constructed of sandstone and brick and be fireproof.[52]

A sandstone and brick city hall would stand on West Washington Street among beautiful and elegant governmental and cultural buildings. The city hall would occupy a site just west of the ponderous red brick and Portage Entry red sandstone United States Custom House, Post Office, Court House, and Government Building (1888–89) by M. E. Bell and Will Freret, supervising architects of the United States Treasury Department, and the Portage Entry red sandstone Marquette Opera House (1890–91) by Demetrius Frederick Charlton, both now demolished. Washington Street was transformed in the 1890s from a thoroughfare of residences into a promenade of masonry public buildings.

The council decided to hold a special election to win the approval of the voters for borrowing fifty thousand dollars and for issuing bonds to pay for the construction of a city hall. Townspeople disagreed over the soundness of constructing a new city hall during the economic recession then in progress. The panic of 1893 had depressed the iron industry, and a scarcity of outside capital for loans had halted building activity. The unemployed met mortgage payments and living expenses with difficulty. This, in turn, slowed down retail business.

The outgoing mayor of Marquette, Sidney Adams, for one, feared spending public monies on the construction of a public building when unemployment was high.[53] Despite the recession and Adams's words of caution, however, the people of Marquette desperately wanted a new city hall. On 11 July 1893 six-sevenths of the voters elected to issue bonds of the city of Marquette for fifty thousand dollars to build a new city hall. Less than three months earlier, Marquette voters had placed Nathan M. Kaufman in its top office. When elected mayor, Kaufman was a man of considerable means, influence, and stature.

In September 1893 the council approved the recommendation of the Committee of Parks, Cemeteries, Public Buildings, and Grounds to accept the plans for the new city hall submitted by architects Andrew W. Lovejoy and Edward Demar of Marquette. Andrew W. Lovejoy had come from Kalamazoo to Marquette to supervise the construction of the prison for Wahlman and Grip. In Kalamazoo, Lovejoy had associated with Alexander Menzie in a firm known as Lovejoy and Menzie, builders.[54] Edward Demar had practiced architecture with E. E. Grip in Ishpeming before associating with Lovejoy. By 1895 Demar worked in a partnership with R. William Gilbert and Demetrius Frederick Charlton, called Charlton, Gilbert and Demar.[55]

On 9 September 1893 the *Marquette Mining Journal* announced that the council had accepted the city hall plans and described them in detail, thus giving the townspeople the first glimpse of the structure:

> The new building will present an appearance when finished that will be a credit to the designers and the citizens of Marquette who so long have felt the need of more satisfactory accommodations for the use of their officials. The building will be three stories and a basement. The general lines will be classical, relieved by several colonial features, particularly that of the main entrance fronting on Washington street. The front will show a stretch of ninety-two feet backed by a depth of seventy-two. The basement and first story will be of Lake Superior sandstone laid up in ashlar work, and the second and third stories will be of pressed brick. A hip roof will surmount the whole and in the center of the front wall will rise a handsome French dome and the flag staff, the roof being finished with iron cresting. The entire outer appearance will suggest massiveness and grace combined.

The newspaper elaborated on the interior. Steps would rise from West Washington Street to the main entrance spanned by a large archway supported by polished granite columns, which, in turn, would lead inside to a wide hall running the full depth of the building. From the main hall of the first floor would open the mayor's office, recorder's office and vaults, police court room, and treasurer's office and vaults. A broad flight of stairs would climb to the second floor. This floor would contain offices for the street commissioner, health officer, city attorney, and city engineer, and the council chamber. The council chamber would rise two full stories with a spacious gallery and decorative ceiling frescoes. The third story would be arranged with a large assembly hall. The basement would hold the police department, which would open directly outside on the west, and a hall, boiler room, coal bins, and store room.[56]

The council issued bonds to build city hall in August 1893 and authorized the controller to advertise at the earliest possible date for bids for construction provided the city hall could be built for fifty thousand dollars. It accepted the full plans and specifications prepared by the architects and authorized advertising construction bids in October 1893.[57] To save costs, the plans called for brick decorated with as much stone as the budget would allow. In December 1893 the council signed a construction contract for $49,983 with Emil Bruce. A building committee was established, and Lovejoy and Demar were appointed superintendents of construction.[58]

Of particular importance was the effort by the council to use local materials and labor in the construction of city hall. The exterior walls of the foundation and first story were Marquette purplish brown sandstone and the upper stories of red pressed brick; the interior was finished in native birch and hardwoods. Both the use of local materials and the construction project itself seemed to be a means of creating work for some of the unemployed.

The building committee instructed Emil Bruce to purchase and use only stone quarried within the limits of the city of Marquette if it was available for the same price as stone acquired outside.[59] The demand for the material for city hall brought employment to the brownstone quarry. The local newspaper noted that James Sinclair, Marquette stone mason, had begun cutting and dressing the stone and added, "It was a very pleasant sound yesterday to hear the merry ring of the anvil in the tool repairing shop and the duller echo of the heavy mallets and hammers as they drove home the wedges that split the big blocks of stone into proper sizes. All day long a crowd of interested idlers hung around the stone cutting shed as though each move made at this point meant a general resumption of activity in our city."[60] As the building went up, the paper noted that "its erection has furnished much employment for her artisans when it was most needed and the structure will stand as a lasting monument to Mayor Kaufman's administration."[61]

A gala occasion marked the laying of the cornerstone of the Marquette City Hall in May 1894. People from Marquette, Negaunee, and Ishpeming celebrated with a parade and dinner. Benjamin O. Pearl, president of the council, called the event one of the most important in Marquette's history, noting that "the public spirit, the public pride, and the public taste, even the commercial enterprise of every metropolis, is measured in a great degree by its public buildings." Pearl spoke of the ability of Lake Superior sandstone to express symbolically the meaning of city hall. "The work of early builders of the city stand[s] on solid foundations of stone chiseled from quarries that underly [*sic*] this imperishable city." Thus, was laid the cornerstone, and this municipal temple dedicated to "the Goddess of Progress and Enterprise."[62]

The simple dedication ceremony for city hall took place at a regular meeting of the common council on 4 February 1895. Mayor Kaufman noted, "Most of the labor which has entered into the building has been performed by Marquette men. The stone was quarried within the limits of this city. The brick were made by Marquette men. Marquette factories have prepared the inside finish, and so on from foundation to roof, outside and inside, the building today stands as a monument to Marquette enterprise and skill."[63] The new city hall must have seemed at the time to deny the pervading economic depression and express a confidence in the iron hills and iron industry.

Above a recessed round-arch entrance, a central mansard hip roof lends civic presence to the three-story, boxlike structure. City officials hoped this monument to the progressive spirit of Marquette's citizens would usher in a new era of commercial and industrial prosperity in the Marquette Iron Range. In 1977 the city sold city hall to a private developer, who adapted it to new use as professional offices, and moved its offices to a new low, flat-roof building, in which only a shallow veneer of variegated Marquette sandstone lines panels of the exterior walls near the main entrance.

Marquette County Courthouse, 1902–4, Charlton and Gilbert; Manning Brothers, landscape architects, Marquette, photograph 1908. (Library of Congress, Detroit Publishing Company Photographic Collection)

Marquette County Courthouse

For the rest of the decade, few sandstone structures were built in Marquette. Not until 1902–4, when a citizens' building committee selected Charlton, Gilbert and Demar's Beaux-Arts Classical design for the Marquette County Courthouse, was another major structure built of Lake Superior red sandstone. In 1902 the voters of Marquette County approved the issuance of bonds worth $120,000 for a new courthouse, and the original wooden Greek Revival courthouse, built in 1857, was moved off the site. In its place, on a hill that slopes gradually toward Marquette Bay, the county commissioners erected the present steel-frame courthouse at a cost of $240,000. Its rock-faced masonry walls, so joyously out of character with its classical detailing, are of the red sandstone of the North Country, a material long out of favor elsewhere. The building is domed, with a three-story central mass flanked by two-story wings; a colossal Doric columned portico marks the entrance. A Doric entablature, the cornice of which is copper, encircles the building. The courtroom on the second floor, which is under the dome with stained-glass lights, is finished with mahogany.[64] At a time when half of the forty thousand residents of the county resided in the iron range towns of Ishpeming and Negaunee, the solid dignified courthouse proclaimed Marquette as the seat of county government. Manning Brothers of Boston were the landscape architects. Under the supervision of Lincoln A. Poley, the courthouse was gloriously restored in 1984.

Peter White Library

The Beaux-Arts Classical design of the Peter White Library, constructed in 1902–4 of smooth-cut grayish white Bedford, Indiana, limestone, contrasts sharply with the warm reddish and purplish brown sandstone and red brick structures in the Gothic, Italianate, and Richardsonian Romanesque styles that were built in the preceding thirty years in Marquette. The library board explained that it chose white stone "so as to furnish a variation to the dark stone which is characteristic of the architecture of Marquette."[65] The *Marquette Daily Mining Journal* for 22 September 1904, argued, "It is of a beautiful whiteness, and as it is the only structure of the material in the city it has a distinctive air that would have been hopelessly lost had Lake Superior sandstone been used." The library stands well back on a spacious lot at the southwest intersection of Front and Ridge Streets. Round-arch windows with keystones pierce the walls on the raised first story; on the second story the windows are rectangular. An ornamental cornice encircles the red tile hipped roof. Stairs ascend to the projecting main central entrance portico, which is marked by four giant fluted columns in antis.

The building was named for Peter White, founder in 1872 of the public library in Marquette and supporter of library programs for thirty years. Together with Nathan M. Kaufman and John M. Longyear, he was also a substantial contributor to the present sixty-thousand-dollar structure. As a member of the World's Columbian Exposition Commission, White probably admired the white classicism of the exposition buildings in Chicago. Patton and Miller, a Chicago firm noted for its designs for libraries and college buildings designed the Marquette library. Today the library is undergoing expansion with a large addition and restoration.

The shift to classicism and whiteness in this major public building, even if only one-third the cost and two-thirds the size of the Marquette County Courthouse, represented a drastic change in taste.

Summary

Situated in the midst of a howling wilderness and harsh climate, Marquette grew as eastern and urban investors from New England, New York, Cleveland, and Lower Michigan speculated in the region's iron and timber resources and its land and developed the city into a major

Peter White Library, 1902–4, Patton and Miller, Marquette, photograph 1905. (Courtesy Library of Congress, Detroit Publishing Company Photographic Collection)

shipping, financial, and commercial center. Immigrant groups, in particular, Cornish, Scandinavian, Scots, and English, provided the labor force. The rush of industrial, business, and commercial activity created the demand for buildings. Marquette achieved architectural unity and distinctiveness because its architects, builders, and quarrymen extensively promoted and used the native Jacobsville sandstone building material. Both professionally trained and untrained architects, contractors, and builders eagerly chose the readily available reddish brown sandstone, and clients and townspeople alike developed a romantic interest in it.

Except for a brief three-year period in the 1870s, no professionally trained architect resided permanently in Marquette until nearly 1890. Architects from Detroit, Chicago, and other urban centers routinely were called upon to design public and private buildings beginning in the 1870s when the city had progressed socially and economically. In 1872 Henry Lord Gay of Chicago drafted plans for the Superior Building. Two years later

Gordon W. Lloyd, Detroit's foremost fashionable architect, designed Saint Paul's Episcopal Church. In 1887 William Scott of Detroit planned the Marquette prison. Gay, Lloyd, and Scott represent those architects who had attended schools, traveled and studied abroad, and worked in the offices of eastern and midwestern cities. These men with established professional reputations were familiar with the high styles in architecture, and, thus, were awarded commissions for Marquette's important public and private buildings. They fulfilled their clients wishes for architecture in the cultivated style. They recognized the uniqueness of the Marquette environment, adapted to the local situation, and freely selected the easily available Jacobsville sandstone for their building material. They exercised restraint in the use of the high style and scaled their projects and designs to conform to Marquette's economic and social conditions. Gay, Lloyd, and Scott guided the fashion for stone and styles in architecture.

Supervision of building projects was frequently assigned to an employee of a Chicago or Detroit firm who opened a branch office in Marquette to oversee construction. At times, it was given to a builder already on location. Thus, Gay dispatched Carl F. Struck to Marquette to superintend the construction of the Superior Building. In turn, the building committee of Saint Paul's Church selected Struck, then living in Marquette, to oversee Lloyd's plans for the new church. And Scott sent Demetrius Frederick Charlton to Marquette to superintend the construction of the prison. Struck and Charlton remained in Marquette, the one for three years, and the other for the rest of his life, where they pursued their careers independently, often working in stone. Both were European-born and had acquired architectural skills in Norway and England, respectively. The contractors and builders to whom the supervision for construction was delegated played key roles in interpreting the design of the architect. Often they were more skilled in the use of stone than the designer.

For the most part, men with little or no professional training built the architecture of Marquette in the 1870s and 1880s. Often they were carpenters, stone masons, or draftsmen who were elevated to contractors and architects by the burgeoning prosperity of the 1880s. Hampson Gregory, a skilled stone mason, applied his experience in handling rock as a miner and apprentice builder in Devonshire to the rocky environment of Marquette. His design and his execution of the Harlow Block and the Merritt house display his innate ability in working with stone.

The group of untrained craftsmen who learned by practice and observation and who responded intuitively and directly to the stone building material due to emerging needs produced a native vernacular tradition in architecture. In this way, Stephen R. Gay, superintendent of the Pioneer Furnace, built stone blast furnaces, George Craig erected the roundhouse

View of Marquette looking northwest. The major civic buildings were built of Jacobsville sandstone or brick trimmed with sandstone. Shown clockwise are the Jacobsville sandstone Marquette County Courthouse; the brick trimmed with sandstone Marquette City Hall; the brick trimmed with sandstone United States Custom House, Post Office, Court House, and Government Building; and the sandstone Frank Carney Block, 1891, Andrew J. Lovejoy, 136 West Baraga Street, photograph 1908. (Courtesy Library of Congress, Detroit Publishing Company Photographic Collection)

and machine shops of the Marquette and Ontonagon Railroad, and mining companies put up utilitarian rubble stone industrial and mining buildings and warehouses.

Easterners and urbanites who managed businesses and industries sought a high degree of sophistication in this wilderness city. Having broader experiences and greater finances that many of the local people, they employed the more highly trained architects and used the purest sandstone. Thus, the clearest, most finely dressed and carved red and brown sandstone was placed in the walls of Saint Paul's Church and the Longyear house. School boards and long-time residents of Marquette chose the variegated or mottled stone because it suited their modest budgets and appealed to their simpler tastes. A regional architecture characterized by the persistence of the vernacular tradition emerged in Marquette. The solid, simple structures are built of the rock they stand on.

By the turn of the century, writers regarded Marquette as the best built, wealthiest, and most beautiful city on the south shore of Lake Superior, yet despite its highly civilized character, the dense wilderness still was never far away. This civilized character can be attributed in large part to its handsome buildings of Jacobsville sandstone. Today Marquette remains the Upper Peninsula's largest city.

Bold red sandstone cliff at Jacobsville, Portage Entry, photograph c. 1985. Over the years nearly seventy companies extracted the famous sandstone from quarries in the Jacobsville formation and the Bayfield group. (Courtesy Balthazar Korab)

Duluth Central High School, 1891–92, Palmer and Hall, Duluth, photograph 1984. Detail of front entrance. The dynamic stone masonry, massive forms, and irregular outlines of the Richardsonian Romanesque style almost seemed to beg for rendition in Lake Superior sandstone. (Kathryn Bishop Eckert)

Upper Peninsula Branch Prison and House of Correction (Marquette Prison), photograph 1988. The High Victorian Romanesque administration building with square central tower and octagonal corner towers is constructed of reddish brown Marquette sandstone trimmed with red Portage Entry sandstone and covered with an immense hipped roof of Lake Superior slate. (Kathryn Bishop Eckert)

Sandstone blocks, Hermit Island, Apostle Islands National Lakeshore, photograph 1984. The Excelsior Brownstone Company operated quarries on Hermit Island. Operations had ceased by 1897. (Kathryn Bishop Eckert)

Bay Furnace, 1869–70, Onota, photograph 1998. Ruins stabilized by the Hiawatha National Forest tell of the operations of the Bay Furnace. Constructed of sandstone from Powell's Point, the furnace fired iron ore, charcoal, and flux to produce pig or cast iron. (Kathryn Bishop Eckert)

Alternating red and white streaked or variegated Portage Entry sandstone. Detail. Small quantities of iron oxides give the rocks of the Jacobsville formation and the Bayfield group their striking red color. The red may be mottled with white streaks. (Courtesy Balthazar Korab)

Marquette County Savings Bank, 1891–92, Barber and Barber, Marquette, photograph 1994. The Rock River sandstone and red brick bank building spoke of the soundness and permanence of this financial institution in the growth and prosperity of Marquette. (Courtesy Tom Buchkoe)

Houghton County Courthouse, photograph c. 1985. The gaudy, polychrome Venetian Gothic creation of Upper Peninsula materials—pale yellow Milwaukee brick (manufactured at Lake Linden) walls, red Jacobsville sandstone trimming, and green oxidized copper roofing—brought the vision of John Ruskin to northern Michigan. (Courtesy Balthazar Korab)

Hancock Town Hall and Fire Hall (Hancock City Hall), 1898–99, Charlton, Gilbert and Demar, Hancock. Thirty years after a disastrous fire in Hancock and as this fireproof Portage Entry red sandstone town hall was built, Hancock enacted a fire ordinance. (Courtesy Sadayoshi Omoto)

Red Jacket Fire Station (Calumet Fire Station), 1898–99, Charles K. Shand, Calumet, Keweenaw National Historical Park, photograph 1997. Evenly colored, reddish brown, top-grade, rock-faced Portage Entry sandstone rises in the south and east walls, and variegated reddish brown and white rubble in the north and west walls. (Kathryn Bishop Eckert)

Washburn State Bank (Washburn Heritage Museum and Cultural Center), 1889–90, Conover and Padley, Washburn, photograph 1998. The recent destruction by fire of buildings in Washburn precipitated the decision to build this bank of Bayfield brownstone from Houghton Point. (Courtesy Mark Fay, Faystrom Photo)

Sainte Anne's Roman Catholic Church (Keweenaw Heritage Center), 1899–1901, Charlton, Gilbert and Demar, Calumet, Keweenaw National Historical Park. This is one of the many great ethnic Portage Entry sandstone churches that gave Calumet the name "the city of churches." (Courtesy Tim Slattery)

Paul P. and Anna L. Roehm House, c. 1895–96, Paul P. Roehm, builder, Laurium, photograph 1998. Paul Roehm, the region's preeminent stone mason and supplier, probably built his own rock-faced Portage Entry red sandstone house himself. (Kathryn Bishop Eckert)

Free Public Library (Washburn Public Library), 1903–5, Henry E. Wildhagen, Washburn, drawing 1903. Front elevation, architect's ink and pen watercolor. Evenly coursed rock-faced light purplish brown Washburn brownstone rises in the exterior walls of this Beaux-Arts Classical Carnegie library, demonstrating the importance of education and culture to the people of Washburn. (Courtesy Washburn Public Library)

Union Passenger Depot of the Wisconsin Central Railroad (The Depot), 1889, Ashland, photograph 1998. The rugged brown Bayfield sandstone railroad station symbolized the hopes of Ashland businessmen for the future of the city as a major shipping center. (Courtesy Mark Fay, Faystrom Photo)

Marquette County Courthouse, 1902–4, Charlton and Gilbert; Manning Brothers, landscape architects, Marquette, photograph 1986. Like the Duluth library, the Beaux-Arts Classical courthouse was constructed of Jacobsville sandstone long after it was out of favor elsewhere. (Kathryn Bishop Eckert)

Duluth Central High School, 1891–92, Palmer and Hall, Duluth. The sandstone architecture of the Lake Superior region culminates at the head of the lake in the great purplish brown Duluth Central High School. (Courtesy Tim Slattery, Grandmaison Studios, Duluth)

3

THE SANDSTONE ARCHITECTURE OF THE COPPER COUNTRY

INTRODUCTION

Projecting for one hundred miles to the northeast, into the broad waters of Lake Superior, the bold thrust of the Keweenaw Peninsula forms the most northerly reach of the state of Michigan. It is a dramatic climax to a varied and beautiful land. There was a time, however, when this remote region was also the repository of one of the state's most valuable resources: a narrow spine called the Copper Range, which runs the entire length of the peninsula, once held rich deposits of copper. This remarkable geological feature is flanked by sandstone.

About forty-five miles from its tip, the Keweenaw Peninsula is pierced by Portage Lake and the Lake Superior Ship Canal, forming the Keweenaw waterway, which opened in 1873. This intrusion into the land virtually transforms a major portion of the peninsula into an island. From the southern shores of Portage Lake, the land rises abruptly to a gentle incline that forms the site for the business district of Houghton. From the northern shores, the land rises more gradually from the village of Hancock to an elevated plateau that extends northeasterly beyond to Keweenaw County. It was this area, known as the Portage Lake Mining District, in particular that at one time contained one of the richest copper fields in the world. At its northern boundary, the villages of Laurium and Calumet, once made up of Red Jacket, Blue Jacket, Yellow Jacket, Limerick, Tamarack, and other enclaves named after mine locations, merge together to form one continuous community.

Prehistoric Indians had taken copper from the Lake Superior region, and French explorers had noted the existence of this red metal as early as the seventeenth century. In 1772 Alexander Henry prospected for copper on the Ontonagon River near Victoria. His *Travels,* published in 1809, reported the presence of a large boulder of copper weighing approximately six thousand pounds near the mouth of the river. Julius Eldred led a well-publicized campaign in 1841 to remove the Ontonagon boulder from the Ontonagon River and display it in Detroit. It was subsequently

Panoramic View of the Keweenaw Peninsula, the Great Copper Country of Northern Michigan. A. K. Cox, Houghton, Mich., 1913. Portage Lake and the Lake Superior Ship Canal pierce the Keweenaw Peninsula. (Courtesy Clarke Historical Library, Central Michigan University)

moved to the Smithsonian Institution in Washington. Further attention was drawn to the area in 1841, when state geologist Douglass Houghton submitted his fourth annual report describing the location and extent of the copper deposits of the Keweenaw Peninsula. Congress purchased the lands from the Chippewa in 1843, and a speculative craze, encouraged by the federal land policy, began and lasted three years. The government abandoned the leasing policy in 1846 because it was difficult to administer. In 1847 it called for an identification of Lake Superior mineral lands and then sold the permits at $2.50 an acre. Later, in 1850, the fee was reduced to $1.25 an acre.

The Copper Country was settled and developed when investors from Boston, New York, Chicago, Marquette, and Lower Michigan speculated in the region's mineral resources and in its rugged land. After prospectors and geologists had identified the location and extent of the mineral resources, Alexander Agassiz, Quincy Adams Shaw, William A. Paine, W. H. Mason, and others established, developed, and finally consolidated the Copper Range, Quincy, and Calumet and Hecla Companies into giant mining operations. Towns grew up around the mines and mills and at the transportation centers of Houghton and Hancock. Hordes of Cornish, Irish, German, French-Canadian, and Scandinavian immigrants, followed by Austrians, Hungarians, and Italians, flocked to the region to work as miners, trammers, and laborers. Mining engineers and technicians arrived from the East to supervise and manage them. The rush of industrial, social, commercial, and educational activity created a demand for buildings. Copper Country architecture achieved a special harmony with the land because its architects and builders used, with sensitivity and understanding, the materials of the area—native Jacobsville sandstone, taken from the site or quarried locally at Portage Entry, and poor mine rock discarded from the mines.

The mining industry was aided by the opening of the locks at Sault Sainte Marie in 1855 and the completion of the Lake Superior Ship Canal in 1873. Railroads also assisted in this development: the Mineral Range Railroad connected Houghton with Hancock in 1873; the Duluth, South Shore, and Atlantic reached Houghton in 1883; and in 1885 a railroad over a drawbridge connected Hancock and Houghton, the major commercial centers of the region.

The architecture that emerged out of the interaction of the land, the resources, and the people in this region ranged from the strictly vernacular for the workers to the fashionable high styles for the owners and managers. The strong vernacular architectural tradition is especially distinguished by the character and quality of Jacobsville red sandstone.

The end of mining came in the twentieth century. Deteriorating labor relations led to the Michigan copper district strike of 1913–14. The Calumet and Hecla reincorporated and consolidated its numerous mining properties in 1923. The depth of the mines increased and the copper content of the ore diminished. The Calumet and Hecla sought other ways to make profits by capturing large quantities of copper lost in the milling process, extracting ore from amygdaloid rock in new mines, and reclaiming copper from mill sands. Mining was hard hit by the Great Depression, and copper mining permanently ended in 1968. In the middle and late twentieth century, tourism, education, and high technology replaced copper as the basis of this region's economy. In 1992 Congress passed Public Law 102-543, establishing Keweenaw National Historical Park in and around Calumet and Hancock.

Houghton

Founded in 1852 and named for Douglass Houghton, Houghton lies on the south shore of Portage Lake. Houghton grew as the chief shipping and distribution point of the Keweenaw Peninsula and the Portage Lake Mining District. Today it is a county governmental, business, and educational center. Docks and warehouses once lined the shore of Portage Lake. Paralleling the waterfront a level above is Shelden Avenue, the main thoroughfare. The old high school, churches, the county courthouse, and houses climb the hillside above Shelden Avenue.

Michigan Mining School (Hubbell Hall), Houghton

The Michigan Mining School was a public building in the Richardsonian Romanesque manner that had special meaning for the people of the

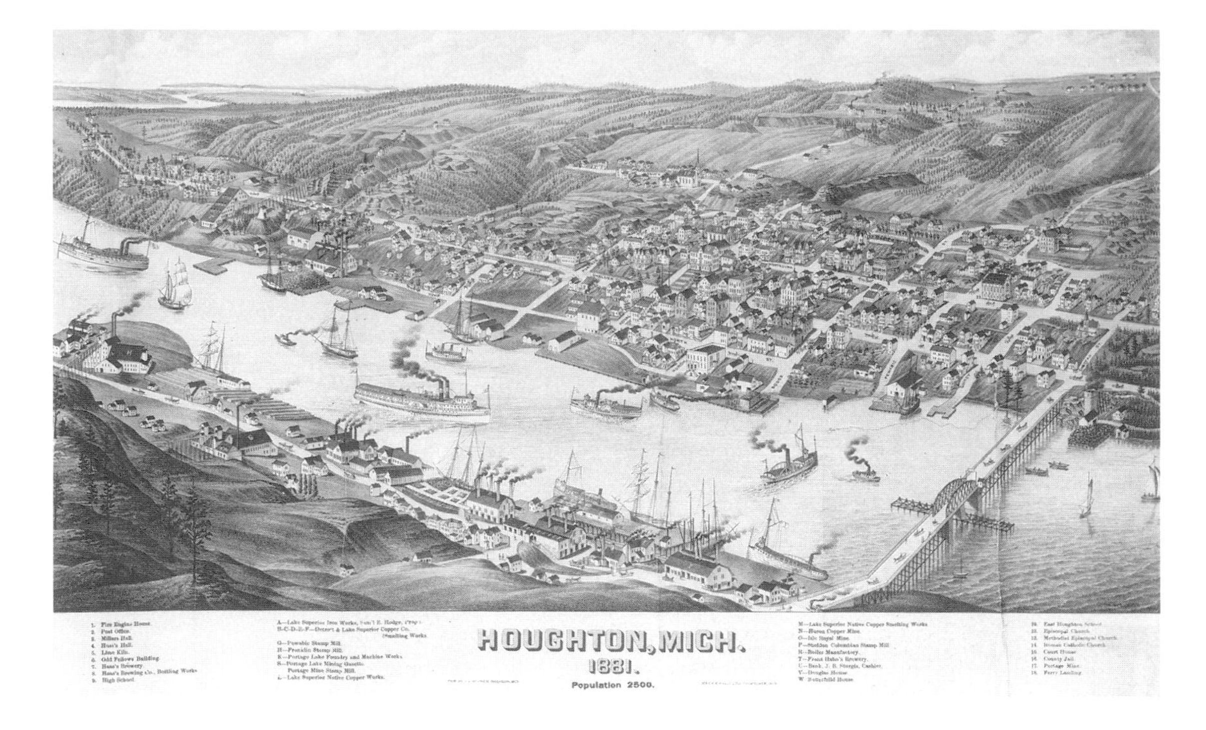

Houghton, Mich., 1881. Beck & Pauli, lithographer, Milwaukee, Wisconsin. A. J. Stoner, Madison, Wisconsin. This bird's eye view of Houghton from Hancock portrays the city before the Portage Entry quarries furnished its builders with red sandstone. (Courtesy State Archives of Michigan)

mineral ranges. Erected between 1887 and 1889, two years after the legislature established the school, the building was situated on a slope above Portage Lake at East Houghton, in the very heart of the mining district. It contained all departments of the school—laboratories, classrooms, offices, and the library. Once named Science Hall, it was later named Hubbell Hall for Jay A. Hubbell, its primary benefactor. Built of Jacobsville red sandstone, the sturdy Richardsonian building gave credence to the importance of the mineral resources of the Upper Peninsula to the state and to the nation.

The Michigan legislature established the Michigan Mining School to prepare men to assist in the development of the country's mineral resources. The school would teach the science, art, and practice of mining, as well as the application of mining machinery, by offering students practical and theoretical training in geology, mineralogy, chemistry, mining, and mining engineering. Act 70 of the (Michigan) Public Acts of 1885 established the school in the Upper Peninsula and appropriated funds for this purpose.[1]

To manage the school, Governor Russell A. Alger appointed a board of men who had participated in the exploration and development of the iron and copper ranges and who knew the requirements of the state's geo-

Jay A. Hubbell (1829–90). From **Memorial Record of the Northern Peninsula of Michigan Illustrated.** *(Chicago: The Lewis Publishing Company, 1895), 381.*

logical and mineral interests. The six-member board of control included James North Wright of Calumet, Thomas L. Chadbourne of Houghton, Alfred Kidder of Marquette, John Senter of Eagle River, Charles A. Cady of Iron Mountain, and John H. Forster of Williamston. Although half of its members died or resigned while the school was established, built, and equipped, men of similar backgrounds were appointed to fill the vacancies. Cady and Forster resigned in 1887, and James N. Wright died in 1888. Charles E. Wright of Marquette and Graham Pope of Houghton were appointed in 1887, John M. Longyear of Marquette in 1888, and Jay A. Hubbell of Houghton in 1889.[2]

Jay A. Hubbell House, 1875, Carl F. Struck, East Houghton, photograph date unknown. The High Victorian wooden house, now destroyed, which overlooked the Portage Canal and the Quincy Mining Company Smelter at Ripley, represented the fine wooden building tradition of the Lake Superior region. (Courtesy Michigan Technological University Archives)

Hubbell was singularly prominent in promoting the school. As a state senator for two terms beginning in 1886, he introduced in the legislature the bill to establish the school, secured appropriations for it, and donated land next to his own large wooden Italianate house built to the designs of Carl F. Struck in 1875 as a building site.[3]

At its first meeting on 15 July 1885 in Houghton, the board of control established the site of the school in or near the village of Houghton. The mines, smelting works, foundries, and rolling mills in the Portage Lake vicinity would furnish students with a practical mining laboratory. For the first few years the school occupied rented space in Houghton's engine house in the Odd Fellow's Building and in a large skating rink building.

Many problems plagued the fledgling mining school. It lacked curriculum, faculty and staff, space, equipment, and reputation. Most discouraging of all, a statewide attitude held that the school was an experiment. In 1887 the board acted to overcome these obstacles and to place the school on solid footing. Only ten days before school was to open for the fall, it appointed Marshman Edward Wadsworth, a Harvard-trained natural and physical scientist knowledgeable in the geology of the Lake Superior region and experienced in developing courses of study, to direct the school.[4] Then the board launched a campaign for a building.

Seeking an appropriation from the legislature for a building, the board set forth its ideas of a structure suitable to house the program of the school and to establish its image. The board employed J. B. Sweatt, a Marquette architect and builder with previous experience working in Chicago, who had designed many public and private buildings in the immediate area. At that time, Sweatt was constructing in Houghton the Houghton County Courthouse, a gaudy, polychromatic, three-story, Venetian Gothic creation of Upper Peninsula materials that brought the vision of John Ruskin to northern Michigan. This is apparent not only in the detailing but especially in the polychrome mixture of pale yellow Milwaukee brick (manufactured at the Seager and Gunnis brick yard at Lake Linden) walls with Upper Peninsula red sandstone trimmings from quarries at Marquette, L'Anse, and Portage Entry, and with the green-oxidized Lake Superior copper roofing. Sweatt prepared preliminary plans for a sandstone mining school building that could accommodate the anticipated number of students and be built in a year for seventy-five thousand dollars. When completed in January 1887, the plans were exhibited at the architect's office, and a reporter for the *Marquette Mining Journal* described them to the public this way:

> The plans are termed "preliminary," but in appearance they were anything but the rough drawings one would expect from the use of that term, being finely executed and of great architectural beauty. They are for a building 152 x 122 feet in extreme dimensions and three stories high, apart from a large basement, the building to be of Lake Superior sandstone, the basement of the Portage Entry red stone in regular courses and the walls of the same stone in broken ashlar, with all the cut stone trimmings of Marquette brownstone. In architectural style the building will partake largely of the renaissance with an adapted Queen Anne roof that style of roof having been found most suitable for this climate. . . .
>
> According to the plans the basement will be 10 feet high and will contain the coal room, boiler room, machine shop, fire assay, and three laboratories with the necessary equipments of vapor chambers, etc., and with one or two large halls. The boiler and engine rooms will have concrete floors and will be fire proof. The first story will be 45 feet high and will contain a carpenter shop, model room, metallurgical room, geolog-

ical shop, museum, director's room, office, central hall, reception room, library, chemist's room, balance room, two baths, janitor's room, closets, laboratories, etc. The second story will be 13 feet high and will contain a gymnasium, spectroscope room, two mathematical recitation rooms, surveyor's office, drawing room with closets, blow pipe room, central hall corridors, storerooms, lavatories, closets, etc. The third story will be 20 feet high and will contain the main lecture hall, hall, workroom and storage rooms and hydraulic elevator will run from the basement to the attic.

> The interior walls will be of brick and wood with iron lath. The floors will be deadened and will be of hardwood, while the interior finish will be of southern and native pine finished in all. The roof will be a truss of wood and iron and the roofing will be copper. The plans for the front windows will be plate and the remainder the best double strength. The plumbing will be of the most modern kind, complete in all its parts; a gas machine put in. The plans provide for all the departments of a well equipped mining school, and the building will be first class in all its appointments and in the manner in which it is situated.[5]

With plans in hand, Hubbell traveled to Lansing and introduced a bill in the legislature to appropriate seventy-five thousand dollars for the erection and equipment of a suitable building for the use of the mining school, including all permanent fixtures, heating and lighting apparatus, and the like. At the same time the board accepted a parcel of land in East Houghton donated by Jay A. Hubbell as the site for the school building.

The board then appointed Charles E. Wright, John Senter, and Graham Pope to a committee to correspond with architects about the plan and design of a school building and to present their findings with conclusions as to the best architect to be employed for the project. On 26 August 1887 the committee reported to the board, and, together with the board, reviewed a written proposal submitted by John Scott and Company of Detroit. Then they asked the firm to furnish drawings and specifications for the school building. Within a month, Scott presented sketches and costs for the proposed school building and signed a contract to complete the plans. The board adopted Scott's final plans and specifications two weeks later.

The summary of the plans for the school was recalled in Wadsworth's and in the treasurer's report of 1890 to the board of control:

> In planning the interior of the building every effort was made to economize the space for recitation rooms and laboratories and reduce the halls to a minimum.
>
> The main building is 109 feet by 53 feet, with a wing 37 feet by 25 feet.

Houghton County Courthouse, 1886–87, J. B. Sweatt, later additions, Houghton, photograph date unknown. (Courtesy Michigan Technological University Archives)

> The basement floor is used for the boiler room, weighing room, machine and workshops, and assaying laboratory. The first floor contains the director's room, reading room, library and laboratories of general and economic geology, petrography and mineralogy. On the second floor are situated the mathematical recitation room, together with the laboratories for physics, mechanical drawing, surveying and mining engineering. The third floor is devoted to the chemical laboratories, chemical lecture room, chemical supply room, balance room, etc.[6]

Later, the *Marquette Mining Journal* enumerated in detail the reasons the Michigan Mining School would become "one of the most prominent scientific schools in America": the fine sandstone school building, together with its location in a mining locale, the credentials of the director and faculty, the favorable attitude of the community, and the resources of the mining industry. The newspaper said, "The rooms, apparatus and general equipment of the mining school will soon be appropriate to the position it is destined to occupy. The new building, to be erected of brownstone next summer, will be perhaps the finest educational structure in Michigan. . . . It will be located on a street as beautiful as North Avenue in Cambridge, Mass."[7]

The board advertised for bids for constructing and heating the new school in the *Detroit Tribune* and the *Marquette Mining Journal* and opened them on 9 November 1887. Five contractors bid on the construction; contracts were awarded to Wahlman and Grip and to I. E. Swift. At this time Wahlman and Grip were executing Scott's plans for several important sandstone buildings in the Upper Peninsula, including the Marquette prison, the First National Bank in Houghton, the Gogebic County Courthouse in Bessemer, and Campbell and Wilkinson's bank in Marquette.

With its knowledge of the geological resources of the Lake Superior region, the board of control consciously chose native Jacobsville sandstone in building an educational institution devoted to training students in the development of mineral resources. The Furst, Jacobs and Company furnished the sandstone for the Michigan Mining School. The *Marquette Mining Journal* for 4 February 1888 noted that the Furst, Jacobs and Company, "the big brownstone firm" of Marquette, Portage Entry, and Chicago, had the contract for supplying the stone, and presumably, for laying it in the walls of the Michigan Mining School at Houghton. The newspaper explained, "Mr. Jacobs has gone to Houghton to commence work on the contract immediately and already the firm has masons on the way to the Copper Country from Chicago and other points." Securing the contract against all competition for what it predicted would be "the finest single building devoted to the educational purposes in Michigan" demonstrated the business ability of Jacobs and his associates.

Workmen excavated the foundation for the mining school as soon as spring arrived and constructed the building in a little over a year. By September 1888 they had laid the stonework on all but the tower and had put on the tile roof, thus enclosing the building before winter. They finished the interior in the winter and spring of 1889, before funds were exhausted.

A second legislative appropriation for sixty thousand dollars in June 1889 enabled the board to furnish and equip the mining school with books, technical equipment, implements and machinery; to landscape the grounds; and to store and bring to it a supply of water.

Finishing and equipping the building took until the spring of 1890. As the installation of the furnishings progressed, the *Houghton Daily Mining Gazette* for 12 December 1889 summarized the advantages of the school

Michigan School of Mines (Hubbell Hall), 1887–89, John Scott and Company, Houghton, photograph c. 1906. Destroyed 1968. (Courtesy Library of Congress, Detroit Publishing Company Collection)

with specific reference to the building: "The building as a whole is a stately structure, worthy of the beneficent institution for which it is used. . . . The mining school will take rank in the near future with the long established public institutions of learning the Michigan people regard with so much pride." The Michigan Mining School advertised its program in the *Michigan State Gazetteer and Business Directory* for 1895–96 with an engraving of Hubbell Hall. The local newspaper cited the distinct advantages of the school's location at Houghton:

> The Michigan copper district is also unequaled as a place in which to imbibe mining knowledge and enthusiasm from her association with the people. Nowhere else in the world are there greater, richer and more successfully managed mines; and nowhere else in America are the people of a large and important district so exclusively devoted to mining

Houghton AND BUSINESS DIRECTORY. Houghton 969

Michigan Mining School

A State School of Mining Engineering,

Giving instruction in Mathematics, Physics, Mechanical Drawing, Designing for Metallurgical, Mining and Mechanical Engineering Plants, etc., Blue Printing, Chemical Analysis, Assaying, Ore Dressing, Metallurgy and Metallurgical Experiments, Properties of Materials, Mechanism, Mechanical and Electrical Engineering, Mining, Gold, Silver, Copper, Lead, Coal, etc., Mine Accounts, Hydraulics, Mineralogy, Petrography, General and Economic Geology and allied subjects. Has two courses, one of three and the other of four years in length. Properly prepared students are admitted at any time. The work extends over some forty-five weeks each year, and the practical work in Surveying, Shop Practice and Field Geology given to the regular classes in the summer, is open to any one qualified to take it. The Laboratories, Shops and Stamp Mill are well equipped. This institution is one of the largest and most practical schools of its kind in the country, as well as the only one devoted exclusively to Mining Engineering. It is situated in a mining region, and the students become familiar with mining practice. Being a State institution, tuition is free to any one, without regard to residence.

For catalogues or information, send to the Director,

M. E. WADSWORTH, Ph. D., Houghton, Mich.

Michigan Mining School advertised its courses with an illustration of its school. From Michigan State Gazetteer & Business Directory *(Detroit, R. L. Polk Co., 1895–96), 969.*

> industries. The profession of scientific mining will grow in importance for many years to come. There is more room in it than in other professions. Success in it requires a vast amount of special knowledge and practice. The Michigan Mining School, being the place above all others where that special preparation can be obtained, will in due time attract its full share of all those seeking such knowledge.[8]

Fully equipped and furnished, the Michigan Mining School opened in 1890. The previous June the State Board of Visitors to the Michigan Mining School had inspected the building. It reported its findings to the Superintendent of Public Instruction, noting the mining school's advantageous location in the midst of the Copper Range and the Menominee, Gogebic, and Marquette Iron Ranges, its competent instructors, its extensive collection of geological specimens, its well-equipped laboratories and library, and its fully developed curriculum. But above all, the visitors found the building "a magnificent one, plain, well built of Portage Entry sandstone, commodious and well adapted in every way to the work of the school."[9]

Scott designed the symmetrical hip-roof structure with a dramatic central tower and round-arch windows. The bold exterior walls proclaimed their purpose: they were laid in random rock-faced ashlar of Portage Entry red sandstone from the Furst, Jacobs and Company's quarry at Portage Entry and trimmed with Marquette brownstone.

Projecting a geological image of the land the building was designed to illuminate, the school not only served but also, through the very nature of its forms, celebrated one of the state's most precious attributes, its rugged spine of mineral resources. The solid red sandstone building affirmed the purpose and function of the Michigan Mining School. The building was demolished in 1968 to make way for an eleven-story mechanical engineering structure.

Shelden-Dee Block, Houghton

Between 1899 and 1901, several Houghton businessmen, who had prospered as speculators in the mineral and timberlands of Michigan's copper frontier, built a group of commercial buildings at the intersection of Shelden Avenue and Isle Royale Street, the heart of Houghton's commercial district. The district served as the commercial center for the Portage Lake Mining District, the most important copper mining and processing area in the Copper Range. The commercial buildings reflect the culmination of Houghton's commercial growth, which was inextricably tied to the mining industry.

The Sheldens, Dees, Douglasses, and Calverlys commissioned Henry Leopold Ottenheimer (1868–1919) of Chicago to design the buildings. Ottenheimer, a student of Adler and Sullivan, was the architect of the Chicago Elks Club and the Chicago Hebrew Institute Gymnasium, among many public buildings, banks, and stores in Chicago; the Temple Jacob (1912) in Hancock; and the Allen Forsyth and Caroline Willard Reese House (1899–1900) at 918 College Avenue in Houghton.[10]

Despite Ottenheimer's association with Louis Sullivan, one of this country's most famous architects, the influence of Sullivan on Ottenheimer's Houghton works is modulated by the strong regional tradition of building with sandstone and by an apparent local taste for classical detail. The Houghton people contracted with Paul F. P. Mueller (1865–1934) of Chicago, an engineer for Adler and Sullivan and Frank Lloyd Wright, to execute the plans prepared by Ottenheimer. The red Portage Entry sandstone buildings—James Dee Block, the Calverly Building, and the huge Shelden-Dee Block—together with the pale yellow brick Douglass House Hotel, show the evolution of the preference for sandstone and brick in Houghton.

The group of structures complemented the prestigious First National Bank, built at 600 Shelden Avenue on the northeast corner of this intersection ten years earlier. The bank, designed by John Scott and Company of Detroit as a three-story Richardsonian Romanesque-inspired work of pressed red brick and Portage Entry sandstone, was the first brick block in the Copper Country. When it was completed in August 1889, the *Portage Lake Mining Gazette* for 15 August 1889 bestowed lavish praise on the structure: "Nothing equal to their brownstone arched and plate glass front, their massive antique oak counter, doors and wood finish, their long line of side windows, and their great floor space and height of ceiling—can be found in any other banking establishment north of Milwaukee." The Douglass House Hotel, the First National Bank, the James Dee Block, and the Shelden-Dee Block still stand; the Calverly Building was demolished to make way for a parking lot.

James R. Dee, manager of the Peninsula Electric Light and Power Company, and Mary E. Shelden (1845–1935) commissioned the large three-story Shelden-Dee Block at 512-514 Shelden Avenue. Dee and Shelden were members of early pioneer families who had speculated in the mineral and timberlands of the copper frontier. Shelden was the widow of George C. Shelden (1842–97), who inherited from his parents, Ransom and Theresa Douglass Sheldon, extensive real estate holdings in Houghton and the adjoining counties in 1878.[11] Seven years after acquiring the building site for the Shelden-Dee Block, George Shelden died.

Red brick and smooth-cut and richly carved Portage Entry red sandstone cover the steel frame of the Sheldon-Dee Block. A number of features

Shelden-Dee Block, 1899, Henry L. Ottenheimer, Houghton, photograph date unknown. (Courtesy Michigan Technological University Archives)

show Louis Sullivan's influence: the tripartite division between base, shaft, and top; the emphasis on the verticals in the engaged pilasters that articulate the frame with recessed spandrels and secondary verticals between; the clearly articulated door motifs; and the ornament in the frieze. A rich and ornate copper cornice supports the roof. The Sheldon-Dee Block is a classical version of the Chicago School design. The large block of stores, office suites, and apartments was divided into identical halves, one for each investor.

The construction of the Douglass House Hotel, now the Douglass House Apartments, at 517 Shelden Avenue in light orangish yellow pressed brick ornamented with pale yellowish white-glazed terra cotta trimming marked the shift away from red sandstone. Even so, red Jacobsville sandstone continued to rise in the exterior walls of a few buildings on Sheldon Avenue in the twentieth century. A case in point is the large, four-story Houghton Masonic Temple Building, now the City Centre, built in c. 1905 by Herman Gundlach to the designs of Charles

W. Maass and Fred A. H. Maass of Laurium and Houghton. The building is at 616-618 Sheldon Avenue. This building is steel skeleton construction with an exterior skin that combines the textures of coursed rusticated red sandstone with smooth-cut red sandstone pilasters.

Hancock

Samuel W. Hill, agent of the Quincy Mining Company, platted Hancock on the north shore of Portage Lake in 1858. The village was named for John Hancock, one of the signers of the Declaration of Independence. On the hill just north of Hancock, the eastern-based Quincy Mining Company had opened a mine and office on land that it acquired from Columbus Christopher Douglass. Douglass had speculated in mineral lands after assisting Douglass Houghton on the scientific survey of the region in 1844.

After Douglass established a store at the townsite in 1859, Hancock began to develop. By 1865 there was a hardware store and an apothecary, as well as banks, churches, saloons, stores, fraternal halls, and boardinghouses, but in 1869 a disastrous fire destroyed three-fourths of the settlement. As a safeguard against future fires, the village was rebuilt with brick and stone structures whenever possible. Hancock incorporated as a village in 1875 but waited thirty years to enact a fire ordinance. The city grew as a commercial and social center for the neighboring residential enclaves that developed on the hill in Franklin Township at the Quincy, Pewabic, Franklin, and Hancock mine shafts. It also served as a waterfront center for copper processing, lumber milling, and shipping. Hancock reached its peak in growth and development with the culmination of productivity at the Quincy Mining Company around 1900.

A strong Finnish, Cornish, and Italian identity still characterizes Hancock's population. The first Finns arrived in 1864, after the Quincy Mining Company recruited miners from northern Europe; they were followed by the Cornish and the Italians.

Finnish Lutheran College and Seminary (Suomi College), Hancock

In the 1890s, just after the opening of the Michigan Mining School, later named Hubbell Hall, the Suomi Synod of the Finnish Lutheran Church established a college and seminary in Hancock a few blocks west from the main business district on Quincy Street. The purpose of the school was to preserve Finnish religion, heritage, and culture, and to minister to the Finnish-American congregations. Juho Kustaa Nikander and other University of Helsinki theologians who made up the consistory had knowledge of the earlier training of Swedish Americans for church work at Augustana College, founded at Rock Island, Illinois, in 1860, and at

Gustavus Adolphus College, established at Saint Peter, Minnesota, in 1862. Thus, three months after the Suomi Synod was organized on 18 December 1889, its leaders called a convention and chose a committee of Finnish-American laymen and theologians to select a site for a Finnish college and seminary.

The Suomi Synod wanted to organize Lutheran congregations in scattered Finnish settlements in Massachusetts, northern Minnesota and Michigan, the Dakotas, and elsewhere. Shortly after its founding, a difference over the authority of the clergy and the authority of the local congregation split the group and led to modifications in the constitution. Nikander led the Suomi Synod and the college from their beginnings until his death in 1919. In that time, the synod grew from 9 to 154 congregations and to 35,262 members, the largest single organization of Finns in America.

While dealing with the establishment of Suomi College, Nikander had to tackle the differences of thought in the Suomi Synod. He carefully separated the two issues. Although recognizing the factions among the Finnish people when it came to religious concerns, he made it clear that there were no factions on the "Finnish college question." Finnish-Americans were united in their support for the establishment of the Finnish college.[12]

Over the course of six years, the Suomi College committee solicited offers and negotiated for a site for the college.[13] The committee approached community organizations, businessmen's clubs, and local governments for prospective sites in places with high Finnish populations. After reviewing prospective sites for the college in West Superior, Wisconsin, Saint Paul, Minnesota; and Marquette and Houghton, Michigan, the Suomi College committee eventually selected a site on Quincy Street in Hancock because of the village's large Finnish population. In the meantime, in 1897, the school opened in temporary quarters in Hancock.

Architects in Marquette, Houghton, and Hancock expressed interest in submitting proposals for Suomi College. The building committee chose C. Archibald Pearce of Hancock as Suomi's architect. Little is known of Pearce's background and training, but he created in 1900 the huge Portage Entry red sandstone Saint Joseph Church in the stamping mill village of Lake Linden at the head of Torch Lake for the French-Canadian congregation, and the Finnish Apostolic Lutheran Church (1897) and the Kerredge Theater (1899) in Hancock. Pearce advertised his architectural office in the Scott's Block at Hancock in the *Marquette Mining Journal* for 1899 and in the *Michigan State Gazetteer and Business Directory* for 1901. He solicited correspondence and stated that he could prepare plans, specifications, and estimates.

On 8 June 1898 the *Copper Country Evening News* reported that plans for the new college building were currently being drafted by architect

Pearce, and soon the building would be started. The cost of erecting the new building was estimated at eight thousand dollars. Within a month the board approved Pearce's plans and drawings with alterations that placed four rooms on the second floor for the college president and accommodated two additional rooms on the first floor by raising interior walls between classrooms. It hired Pearce to superintend the building project and sought bids from contractors. In August and September the ground was excavated by John Paavola, and in October a contract to lay the foundation was awarded to William Scott of Hancock. But the board delayed the erection of the superstructure until Pearce could simplify the plans so as to reduce costs.

On 10 March 1899, when the board of directors called for construction bids, the *Marquette Mining Journal* described the plans for Suomi College then on display at the office of C. Archibald Pearce: "The site is a most commanding one on Quincy street, west. . . . The plans call for a building 76 by 63 feet 3 inches in extreme dimensions, while the main portion, excluding the tower and president's office in the front, will be 76 by 40. Rock-faced Portage Entry sandstone laid at random, with cut stone trimmings, will be used in the construction with cut stone for the basement, and the building, Gothic in design, will be a notable addition to the public edifices of Hancock and the copper country." The front portion of the basement would contain the dining room, pantry, and kitchen for the president. The main portion of the basement would hold a classroom, the main dining room, pantry, kitchen, laundry, and fuel and boiler rooms. The first floor, divided by a large hallway that ran from front to rear, would hold the president's reception hall and six classrooms, two of which could be opened by means of a sliding partition into a large public assembly room with a handsome fireplace. The second and third floors would contain student bedrooms and bathrooms, and the attic the infirmary. The upper three floors of the tower would hold the president's parlor, bedrooms, bathroom, and library. The interior would be treated simply; the rooms of the basement would be finished with oiled Norway pine wainscotting, the hallways decorated with burlap wainscotting, and the rooms of the upper floors painted.

By the end of March, the building committee awarded the contract for carpentry work to Bajari and Ulseth of Calumet and for stonework to William Scott of Hancock. Within two months the cornerstone was laid.

On Decoration Day 1899 hundreds of Finns traveled to Hancock to participate in the day-long festival to observe the laying of the cornerstone. Special trains carried delegates from Calumet and from the towns of the iron range. Two hundred people ate dinner at Germania Hall, and then, led by the Laurium band, paraded through the main streets of Hancock to the building site where thousands had gathered to observe the

Finnish Lutheran College and Seminary (Suomi College), 1898–1900, C. Archibald Pearce, Hancock, photograph 1905. (Courtesy Library of Congress, Detroit Publishing Company Photographic Collection)

ceremonies. Finnish theologians addressed the crowd in Finnish. Archibald J. Scott, mayor of Hancock, concluded the speeches by saying that "Marquette has got the normal school and Newberry the asylum, but Hancock has got the Finnish Theological college."[14]

In an article titled "Gala Day for the Finns: Cornerstone of Suomi College at Hancock is Laid with Appropriate Exercises," the *Marquette Mining Journal* for 3 June 1899 concluded:

> Besides being an ornament to the village, Suomi College will be a credit to the educational institutions of both Hancock and the county. The site is admirable, in fact, a more desirable piece of property could not be had along Portage Lake anywhere. The structure will be of Portage Entry sandstone, of liberal dimensions, three stories in height,

handsomely trimmed and most attractively and conveniently arranged throughout. . . . The Finnish people have every reason to be proud of their undertaking in the direction of higher education, but the work now so well underway was not accomplished without sacrifice. It is such events as these that exhibit the traits of a people as well as exciting the sympathy of all lovers of liberty should their kinsmen in their native land be under the heel of oppressors, as is the case of the Finns, who have flatly refused to yield to the tyranny of the czar of Russia, who, by the way, is first in peace conferences and last to consider the rights of his unfortunate subjects.

As the school neared completion in December 1899, the local newspaper illustrated "the imposing structure" and explained that "the style of architecture is Old English Gothic with eliptical [*sic*] Gothic arches. . . . The walls are built of Portage Entry red sandstone heavily buttressed."[15]

The school cost thirty thousand dollars to build. It was ready for occupancy by mid-December and dedicated on 21 January 1900. Classes assembled for the first time on 22 January. The *Houghton Daily Mining Gazette* outlined plans for the dedication which would serve to congratulate the Finnish people on this great undertaking. "The handsome brownstone institution . . . will be an ornament on Quincy Street and a credit to the city for all time. Such a imposing structure, for a purpose so noble and elevating commands the admiration of every resident of the copper country, and of Hancock particularly, and the Finnish people are entitled to the highest need of praise it is possible to bestow for the accomplishment of their plan and the bright promise that the future affords them for the success of their ambitious project."[16] Some six hundred visitors from the Copper and Iron Country towns of Michigan and Minnesota attended the dedication exercises. The president of the college, ministers of several Finnish Evangelical Lutheran churches in Minnesota and Michigan, and the editor of the *American Suometar,* a Finnish newspaper, spoke. The *Houghton Daily Mining Gazette* observed that for a day and a half the city of Hancock was practically in the possession of the Finnish people who feasted, sang, and attended services. "The event was one of extraordinary import to that nationality, as it is the only Finnish college in the United States and enters upon its career under the most auspicious circumstances."[17] The *Houghton Daily Mining Gazette* for 20 January 1900 reported that the dedication would celebrate the founding of the Finnish people's first educational institution of its kind in America soil. The school, the newspaper said, will forever stand as "a monument to the enterprise, energy, and thrift of that race. The building is a marvel of completeness and architectural skill . . . a credit to the architect, the builders, and the society."

The first building at the Finnish-American college is presently known as Old Main. With its rock-faced reddish brown Portage Entry sandstone walls, this rugged and somewhat crudely Richardsonian building is reminiscent of a medieval castle. In its brooding boldness, robustness, and fortresslike appearance, it conveys an appropriate sense of great strength and tenacity. Altogether, it symbolizes the unity of the Suomi Synod, the endurance of Finnish people against displacement by Czarist Russia, and the validity of Finnish culture, tradition, and values in the midst of the American melting pot. Although the building once housed the president and provided a dining room, classrooms, and assembly rooms, today it is a women's dormitory.

As soon as the Suomi Consistory decided to establish the college, the board of directors began raising funds for the project. By July 1895 it had appointed fund raisers who canvassed Finnish settlements throughout the country. Within three years they raised seven thousand dollars. On 12 April 1898 the board instructed the building committee to circulate fund-raising letters to friends of the college, people acquainted with the project, and the public. Mining companies, public officials, and Finns in Finnish settlements throughout America contributed funds for the school.

After the school was occupied, the treasurer reported a debt of twenty-two thousand dollars on the building and grounds, so fund raising expanded to reach great American philanthropists, presidents of mining companies, public officials, and newspapermen who could publicize the cause.[18]

J. H. Jasberg, business manager for the committee, sent this letter to John D. Rockefeller, Andrew Carnegie, Helen Gould, Henry C. Frick, John Wannamaker, William A. Paine, A. S. Bigelow, and others. It conveys his expression of the good that would come from supporting the college.

> By examining the records of the Bureau of Immigration, it will be found that, a very small percentage of the Finnish immigrants are unable to read or write their own language, and they are all Evangelical Lutherans.
>
> The encroachments of the Government of Russia upon the rights and privileges of the people of Finland have become most severe and far-reaching during the past two years than ever before, and have caused a considerable increase in the immigration of Finnish subjects to the United States, the great majority of whom are between the age of 20–25 years. All of them have reached an age where the acquisition of a foreign tongue with the means at their command, becomes well-neigh [*sic*] impossible. If these people were dependent upon their knowledge of the English language for their enlightenment as to public affairs in this country, they would undoubtedly make a very poor class of citizens, and their attendance of American churches, considering they would not comprehend the teachings, would be reduced to a minimum.

> Under the circumstances there is almost a public necessity for school's like ours, as the pulpits of the various churches and the editorial chairs of the Finnish publications in the United States will have to be filled with men educated in our language for a good many years to come.[19]

The influx of some two hundred thousand Finns to America was caused largely by the attitude of the Russians toward Finland. The Finnish-American population of Houghton County followed closely with deep interest the policies and practices of the czar. Discussions of political conditions in Finland drew large crowds in Calumet and Hancock.

In the twentieth century, Suomi College expanded and diversified. A seminary was added in 1904, and the school became a junior college in 1924. The seminary was moved to Illinois in 1958, but Suomi remains the only institution of higher education to be established by Finnish Americans in the United States.

Hancock Town Hall and Fire Hall (Hancock City Hall), Hancock

One block east of Suomi College on Quincy Street opposite Montezuma Park and overlooking the Portage Canal stands the Hancock Town Hall and Fire Hall. In 1897, as the population of Hancock reached four thousand, doubling in the previous ten years, the people of Hancock decided they needed a new village hall and fire hall. They had outgrown their old wooden fire house, built at the east end of Quincy Street in 1871, two years after a fire had destroyed three-fourths of the village. The fire house was too far distant from properties in the new west end of town, too small to house a team of horses and fire-fighting apparatus, and, above all, too inconspicuous in its location and appearance to proclaim the social and economic advancement and the independence of the village. Citizens and city officials alike wanted a proud, modern, red sandstone public building that would house the fire department, village office, and council room and that would visibly proclaim the achievements of Hancock.

Responding to the practical and social needs of the community, in February 1898, the Common Council of the Village of Hancock acquired a centrally located site from the Quincy Mining Company for twenty-five hundred dollars. Seventy-five feet in width by 120 feet in depth, the lot was situated on the north side of Quincy Street, where its eastern boundary was in a line with the east side of Montezuma Street. Then the council relinquished its lease to the land on which the old fire hall was situated, with the understanding the village could occupy it for an indefinite period. The council authorized its president, Archibald J. Scott, to receive plans and specifications for a town hall and engine house at a cost of about ten thousand dollars.[20]

At least three firms submitted plans for the proposed fire hall. William T. Pryor of Houghton, C. Archibald Pearce of Dollar Bay and Hancock, and Charlton, Gilbert and Demar of Marquette and Milwaukee presented plans to the council in a special session held to review them. The council dismissed those prepared by Pryor since they were incomplete, having been drafted on a day's notice. It deferred deciding between those prepared by the local firm and the out-of-town firm. The local newspaper reported that popular opinion favored giving the job to a local architect.

Pearce had drawn a sketch for the proposed village building. It seemed an imposing structure with a single high tower in which to hang and dry the long fire hose and with a conveniently arranged interior.[21] Charlton, Gilbert and Demar submitted plans, and the *Copper Country Evening News* for 11 February 1898 elaborated on them in this way:

> The plans furnished by the Milwaukee firm were highly colored and took the eye of most of the councilmen at once, regardless of giving any attention to the convenience of the structure. The arrangement of the second floor of the plans pleased all the councilmen and filled the bill exactly, but to get this arrangement the designers had to go out of the exact size for which the plans were asked. Plans were asked for a building 40 x 90 feet, but these plans call for a building 48 x 90. The lower floor did not give much satisfaction, the only entrance to which is through the large swinging doors in front. The exterior of the building with its gay colors is pleasing to the eye in the extreme, but as far as appearance and convenience is considered Mr. Pearce's plans are far ahead of the others. While the Milwaukee firm's plans fill the bill to a percentage of 23 percent, Mr. Pearce's plans have a strong 75 percent in its favor to what is wanted. His plans show an imposing structure with a pleasing exterior. The interior being fitted up very conveniently and with but a few changes is just what is wanted.

The council deliberated in two more sessions before deciding on 23 February 1898 to adopt the plans and specifications of Charlton, Gilbert and Demar. The lengthy and heated deliberations of the council probably were influenced by Archibald J. Scott, the local authority on fire hall design.[22]

Edward Demar probably played the principal role in designing Hancock's town and fire hall. Born in Vermont in 1864, the Toronto-trained architect had been employed as a draftsman in several Canadian and Upper Peninsula offices before forming a partnership with Andrew Lovejoy by 1892 and designing the Marquette City Hall with Lovejoy in 1894. In 1899, four years after joining Demetrius Frederick Charlton and R. William Gilbert in the firm of Charlton, Gilbert and Demar, Demar worked in the firm's Milwaukee office for two years. He had married Kate Hoffenbacher, daughter of a Hancock baker, in 1890. Eventually Demar opened his own office at Sault Sainte Marie, where he designed several important sandstone structures, among them, the Adams Block (c. 1902).[23]

Archibald J. Scott was a banker, developer, community leader, and public official. Throughout his entire career, his special interest was fire fighting. He served as chief of the Hancock Fire Department for more than twenty-five years after it was organized in 1870, established a paid company of twenty men in 1882, and was president of the Upper Peninsula Firemen's Association. Coming to the Upper Peninsula from Wisconsin and the West in 1865 at age seventeen, Scott participated in the growth and development of Hancock: he built more than twenty houses, the Scott Hotel (1880), and other structures; operated a drug store; started a local newspaper; and served as vice president of the First National Bank, director of the Superior Trust Company, president of the Eva Mining Company, director of the Hancock Loan, Mortgage and Insurance Company, and mayor of Hancock.[24]

For good reason the townspeople valued Scott's thoughts on the best design for the village hall and fire hall and his energy in bringing them to execution. Scott had visited hundreds of city fire halls throughout the state in search of information on the best building for Hancock. To assist him in taking charge of the construction of the town hall, Scott appointed a building committee. It included Charles E. Wright, state geologist, and William Kerredge, a local businessman. Wright was later replaced by W. H. Mason, an employee of the Quincy Mining Company.

Although the townspeople had hoped that the fire hall would be ready in time for the Fireman's Tournament the first of August, the structure was not completed until the following winter. The newspaper blamed the out-of-town firm for the delay.

Bids for construction of the foundation were advertised on 11 April 1898.[25] In an effort to move the project along, the council authorized William Scott, a stone cutter in Hancock, to build the stone foundation for the new city hall even before detailed plans of the structure were in hand. Scott completed work on the basement as the full plans were finished and contractors reviewed them for bids. Two weeks later E. E. Grip and Company of Ishpeming was awarded the construction contract. Stone for the city hall was transported by scows through the Portage Canal from the quarries at Portage Entry to Hancock. At the construction site stonecutters prepared the stone for seating in the walls. As the almost ninety-foot-high tower neared completion at the end of September, observers awaited the installation of the clock. Grip completed his work to the full satisfaction of all that fall.[26] Associated Artists of Milwaukee decorated the assembly hall and council chamber. Angus Gillis painted the building, and

John Funke put in the steam heat. Finally, the council ordered furniture for its own chamber from the Lower Peninsula. The entire project cost fifteen thousand dollars.[27]

As workmen cleaned up the structure, and as officials moved in and prepared for the formal opening, the local newspaper described in detail the town hall and fire hall:

> The building, of Portage Entry sandstone, is two stories high, with a Venetian clock tower at the corner by the main entrance, and is 49 by 96 in ground dimensions. The front elevation, with its tower and the battlements and the large triple window in the center, presents a splendid appearance from the street and is a credit to the architects, Charlton, Gilbert & Demar of Marquette and Milwaukee.
>
> The first floor is given up to the fire department, with the marshal's office on the east side with a separate entrance, the "cooler" in the basement also being separated from the main basement. In the fire department's headquarters the new hose sleigh, just finished by Paul Exley and a beautiful piece of work, by the way, occupies the place of honor, while in the rear are the snow safety stalls for the team, a room for two of the firemen and a room for the driver. Hardwood floors, pine wainscotting, natural finish, and 16-foot ceilings make the quarters the handsomest of their kind on the peninsula.
>
> At the head of the stairs on the second floor is the clerk's office, with a fireproof vault, and occupying the front of the building is the council chamber, 20 by 30, with a recessed alcove for the president and convenient cloak rooms. In the rear is the firemen's assembly hall, 30 by 60 feet in size, and the handsomest feature of the building. This room, with its hardwood floor, vaulted ceiling and heavy cornices, has been decorated in warm terra cotta, shading into lighter colors in the ceiling and elaborately frescoed. It is a masterpiece in its way, and with the council chamber reflects the greatest credit upon the Associated Artists. The assembly hall is lighted by 120 incandescents set in the ceiling.
>
> A clock tower contains a large Seth Thomas town clock which has kept perfect time from the day of its installation. The bell will be used as a fire signal and also to strike the hours, the Gamewill system of fire alarm telegraph being used.
>
> Taken all in all the building is a great credit to Hancock, to the present village officials and to the architects, Messrs. Charlton, Gilbert & Demar.[28]

The Hancock Town Hall and Fire Hall was built of stone in compliance with a fire ordinance enacted while it was under construction. On 4 October 1899, as the masonry local government building was readied for

Hancock Town Hall and Fire Hall (Hancock City Hall), 1898–99, Charlton, Gilbert and Demar, Hancock, photograph c. 1900. (Courtesy Michigan Technological University Archives)

occupancy and thirty years after a fire destroyed three-fourths of the town, the village council adopted an ordinance that prescribed fire limits and required persons about to construct buildings within these fire limits to apply for a permit. The ordinance prohibited the erection, placement, or enlargement of wooden buildings and structures in the densely populated center of the village. The ordinance required that buildings erected, placed, or enlarged within these limits be constructed with exterior walls of stone or brick, roofs covered with slate, metal, gravel, or other noncombustible material, and windows and other openings provided with fire shutters.[29]

Over two thousand people inspected the new city building when it was opened to the public on the evening of 16 February 1899. Accompanied by the Quincy Excelsior Band, the village fathers and the fire company marched through the streets from the old fire house to the new town hall. Scott presented the building to the citizens of Hancock as he addressed them from atop the hose wagon saying, "The Calumets and Baltics can come and go, but Hancock would go on forever."[30]

The Hancock Town Hall and Fire Hall shows the Dutch stepped and Flemish curved gable in its design. It was constructed of evenly coursed, rock-faced, clear reddish brown Portage Entry sandstone in a manner reminiscent of H. H. Richardson. The two-story, end-gable building with a broad-arch front window has a square clock tower at one side, above which was originally set a tall spire. The building housed the fire department on the first floor and the village offices and council chambers on the second.

The structure was built as a stable and ornamental local government building during the years when the local mining industry achieved one of its greatest periods of growth. The solid town hall symbolized the stability, security, and permanence of Hancock in a region of many transitory and impermanent mining towns. Under the direction of Francis J. Rutz of Hitch, Incorporated, in 1984–85 the city hall was restored and rehabilitated, and an elevator was added.

Calumet (Red Jacket)

Platted in 1868 and incorporated in 1875, the village of Red Jacket, named Calumet in 1929, grew up on the northwest edge of the Calumet and Hecla Mine. The Calumet and Hecla Mining Company did not permit businesses on its property and had no company store, so stores and saloons were built beyond company boundaries, together with social halls, government buildings, and churches. From its business district along Fifth and Sixth Streets, Red Jacket with a population of 3,078 at the turn of the century, served a larger mining community of 40,000. Today about 7,000 people live in the Calumet area.

Red Jacket Fire Station (Calumet Fire Station), Calumet

Situated on Sixth and Elm Streets in the main business district of Calumet, formerly Red Jacket, are the Red Jacket Fire Station and the Red Jacket Town Hall and Opera House. The Common Council of the Village of Red Jacket, under the leadership of John R. Ryan, the council president, launched a campaign to build a new opera house under the guise of building and improving its public buildings. To accomplish this, the fire department was moved from the village hall across Sixth Street to a new structure designed and built in 1898–99 to the designs of Charles K. Shand.

The fire department and village offices formerly occupied a two-story brick and brown Marquette sandstone structure on the site at the southeast corner of Sixth and Elm Streets that the Calumet and Hecla Mining Company leased to the village. The first floor contained the village clerk's office, the council chambers, the fire hall, and the jail. The second floor held a public hall and opera house. J. B. Sweatt created plans for the structure in 1885, and Jesse Butler built it for $14,272 in 1886.[31]

Red Jacket continuously improved its fire protection system: in 1887 it organized a company of paid rather that volunteer firemen, and in 1897 it installed fire alarms in the homes of every member of the fire department.[32]

On 6 April 1898, at the first meeting of the newly elected village council, presided over by John R. Ryan, the council discussed building a new fire engine house and enlarging the opera house so as to increase the seating capacity of the auditorium, strengthen the facility for theatrical purposes, and improve the town hall.

A committee comprising President Ryan, Aldermen Knivel, Lisa, and Gribble, and Clerk Ellis was appointed to confer with Alexander Agassiz, manager of the Calumet and Hecla Mining Company, about erecting a more efficient fire station on the vacant lot opposite the village offices and opera house on Sixth Street and about adding to or enlarging the opera house.

Agassiz surely supported the program to improve fire protection, for on 26 May 1898 the *Copper Country Evening News* reported that the village council reviewed the plans for the new engine house prepared by four architectural firms: Charlton and Demar, Donald M. Scott, Charles K. Shand, and Charles W. Maass. They selected those of Charles K. Shand.[33]

Shand was an itinerant architect who arrived in Calumet to participate in the building rush of the late 1890s. He advertised his architectural services in the *Calumet, Houghton, and Hancock City Directory* from 1895 to 1904. First alone, and later in association with George D. Eastman, a draftsman who ran a branch office in Calumet for Saginaw architect Clarence L. Cowles, he worked in Calumet, Laurium, Lake Linden, and Houghton and designed several commercial blocks, banks, fraternal halls, town halls, and at least one church and one house.

With plans and specifications complete, the council awarded contracts to Procissi and Company for stone mason work and to Bajari and Ulseth for the carpentry and materials. By mid-August work on the structure began under Shand's supervision. Subsequently, the council awarded to Frank J. Zoberlin a contract for the heating work and materials, and to John F. D. Smith a contract for the plumbing work. The local newspaper predicted the fire station would be the finest and most artistic, convenient, and complete in Michigan and the Northwest and that neither Detroit nor Duluth could boast of such a hall.[34]

The stone masons completed their work on the fire hall by mid-November, and other contractors finished that winter and spring. Early in May the decorators and painters put on the finishing touches, the electricians finished the wiring, and the council prepared to accept the fire station. Construction was complete later in May. The fire station cost just over twenty thousand dollars.

The two-story Richardsonian Romanesque Red Jacket Fire Station has a low, open campanile with a pyramidal roof that served as a hose drying tower at the southeast corner, balanced by projecting pedimented parapets at the center and the northeast corner. A prominent arcade at the first level, marked by three large arched openings for the firehouse doors and by flanking smaller pedestrian entry arches at the corner, dominates the facade. A stepped-gable parapet directly above the middle arched opening attempts to give the work a degree of formality. Evenly colored, reddish brown, top-grade, rock-faced Portage Entry sandstone rises in the south and east walls, and variegated reddish brown and white rubble in the north and west walls.

The firefighting equipment was housed on the first level, and fireman's quarters on the second level. The apparatus and horses occupied the first floor and the basement, each easily accessible to the street through large doors at the front and rear of the building. The basement was divided into a boiler and storage room for trucks and village-owned vehicles, and the first floor contained the engine room, with one box and eight single horse stalls and a harness room. Fireman lived on the second floor in comfortable quarters that included a large recreation room, a reading room, a sleeping apartment for twelve, a bathroom, and the marshal's office. From this floor, hay and feed were dropped to the horse stalls below. The building was steam-heated throughout, lighted by electricity, and supplied with plumbing. Today the fire station serves as a firefighters' museum.

Red Jacket Town Hall and Opera House (Calumet Village Hall and Calumet Theater), Calumet

On 26 November 1898, with construction of the fire station underway, Red Jacket voters elected to borrow twenty-five thousand dollars ostensibly to enlarge their town hall. In reality, the village councilmen sought to erect a new opera house. For more than a year, the village council and taxpayers had debated the issue of building new village offices and an opera house. At regular meetings of the common council on 2 November and 15 November 1898, councilmen summarized their views on the matter.

Alderman Michael Kemp maintained that an addition to the village hall on the city lot adjoining the existing building would provide a large, comfortable auditorium and proper offices for village officials. Kemp urged the council to advertise for plans and initiate steps immediately, arguing that the village must assume the responsibility for the construction of the much needed opera house since the Calumet and Hecla Mining Company would lease the lot to the village but require a private party or corporation to purchase a site. Kemp added that businessmen favored the project. Alderman E. R. Miller and several citizens thought the improvements were necessary if Red Jacket was to surpass Laurium. Alderman Joseph Vertin suggested that the council proceed with plans to enlarge the village hall but let the voters decide on the opera house project. Mayor John R. Ryan concluded that, despite the council's authority to decide the issue itself, a vote was prudent when spending twenty-five thousand dollars of taxpayers' money. The council agreed and passed a resolution to hold a special election on 26 November 1898 to decide whether the village should borrow the money to enlarge the present town hall.

Some people questioned the soundness of erecting a new opera house when the existing one had been refurbished recently. In 1896 the building had been enlarged and redecorated, and new stage lights and opera chairs had been installed. But voters approved the proposal, and the council set about preparing to give the city what the local newspaper termed "a good building and an opera house that would be a credit to the town."[35] In reporting the election results, the newspaper discussed the primary objective of the building program.

> It is generally understood that the principal part of the improvement will be in making a modern opera house out of the present building, as well as building proper offices for the village officers and the common council. . . . The opera house will probably be in the second floor of the addition while the entire first floor will be for municipal offices. The new opera house will be a modern one in every particular and large enough to accommodate theater-going people of the community. One point in

the building that will be particularly well taken care of will be the stage, which will be sufficiently large to allow any company to place the most elaborate scenery upon it.[36]

Five architectural firms, three with reputations beyond Calumet, responded to the village's immediate call for proposals to enlarge the present town hall by submitting plans to the council on 3 January 1899. After reducing the choice to those drawn by Charles K. Shand and by Charlton, Gilbert and Demar of Marquette and Milwaukee, the council selected Shand.[37]

Shand proposed an addition that would more than double the size of the existing hall. Reworking the north and west facades of the existing structure would unify the old and new towered sandstone composition. No changes were planned for the second floor opera house, but the space would serve as a hall for dances, public meetings, and small theatrical productions. The fire department quarters would be converted into a kitchen and dining room, the council room and jail would be enlarged, the village clerk provided with an ample office, and the vaults moved.

Attention focused on the opera house. From Sixth Street a large triple entrance would lead through a vestibule to an aisle at the back of the orchestra seats. The lower floor would hold the saucer-shaped auditorium with seats that sloped toward the stage. The stage would rise sixty feet to accommodate dropped scenery, its proscenium extending thirty-five feet. In fact, the local newspaper predicted the stage would be "the largest and best appointed this side of Milwaukee."[38] The main hall, four boxes, and balcony would seat twelve hundred. Ample dressing rooms would be backstage.

In response to citizen criticism over the lack of consideration for town offices, Shand revised his plans. The council asked the architect to reduce the size of the dining room and to provide an office for the marshal, sleeping apartments for the janitor, and a committee room adjoining the council room, and to move the jail from the basement to the first floor. The *Copper Country Evening News* stated that these revisions give "the town building more the air of a city hall than of an Opera House."[39]

When awarded the commission to plan the town hall and opera house, Shand engaged Byron Pierce, a Hancock architect and builder, to assist him with his projects. Shand then traveled to Chicago, Detroit, and other cities to study opera houses and stages. In fact, having done the fire station and the town hall and opera house, Shand soon advertised public works as his specialty and illustrated the ad in the city directory with a sketch of the Red Jacket Town Hall and Opera House.

On 11 March 1899 the council appointed Shand to supervise construction of the town hall and opera house. It awarded the major contracts

Red Jacket Town Hall and Opera House (Calumet Village Hall and Calumet Theater), 1899–1900, Charles K. Shand. Calumet, Keweenaw National Historical Park, photograph date unknown. (Courtesy Michigan Technological University Archives)

to the following: Paul P. Roehm, Calumet, stone and brick masonry and excavation; Bajari and Ulseth, Calumet, carpentry, lumber, millwork; Frank B. Lyon, plumbing and heating; Keelyn and Smith, Milwaukee, wiring and electrical fixtures; William Eckart and Company, Chicago, decorating; Heywood Brothers and Wakefield Company, opera house seating; associated Artists of Milwaukee, scenery; and John F. D. Smith, plumbing.[40] The council initially had anticipated that the project, including furnishings, would cost no more than thirty thousand dollars, but it cost more than twice this amount.

When the building was completed and open for the first time to public inspection, Shand received praise for his work. The local newspaper called the theater and city hall "the finest north of the Straits of Mackinac" and elaborated, "The front is of finely cut Portage Entry sandstone for the most part. The style of architecture is the Italian renaissance. Cream-colored hydraulic-pressed brick from Illinois forms part of the front. The cornices will all be of copper and the top of the tower will also be of copper, so that in the copper and sandstone, copper country products form a conspicuous part of the building."[41]

Indeed, the light yellowish brown brick town hall and opera house rests on a first story of Portage Entry sandstone. The triple and paired round-headed openings, blind arcades, and a balustrade suggest Renaissance features. The whole is covered with a copper roof and trimmed with copper cornices. A square clock tower at the northwest corner of the opera house structure originally rose to a square open bell tower, which in turn was surmounted by an octagonal cupola.

The theater opened on 21 March 1900 to an audience of Copper Country society with a Broadway opera company performing *The Highwayman*. The local newspaper recorded the event: "The only municipal theater in the United States of America was opened last evening in a blaze of glory when the new Calumet theater built and furnished by the village of Red Jacket was the scene of the first-night performance. . . . The opening was a big success. . . . The new theater in its complete entirety is something exceptionally elaborate for this section of the country."[42]

Before the curtain went up, John R. Ryan, president of the common council, called upon William R. Parnall, superintendent of the Bigelow mines. Parnall spoke of the lasting benefit of the theater to the city and expressed his disbelief that the theater in its appointments could have been built for the price.[43]

The material, style, substance, and use of this public building for cultural purposes affirmed Red Jacket's importance in the Lake Superior region. The building continues to serve in its original use and is a centerpiece of the Keweenaw National Historical Park.

Sainte Anne's Roman Catholic Church (Keweenaw Heritage Center), Calumet Township

French-Canadian Catholics, members of the Saint Louis Roman Catholic congregation, built in 1900 a red Portage Entry sandstone Gothic church at the southwest corner of Scott and Fifth Streets in Calumet. Sainte Anne's Roman Catholic Church firmly and visibly separates the residential, commercial, and civic from the industrial areas of Calumet.

Sainte Anne's Church replaced a hall that had been converted into Saint Louis Church on its founding by French-Canadian Catholics in 1884, one year after the Marquette Diocese had dispatched the Reverend Vermer to oversee the parish. As the mining companies sought the skills of French Canadians as suppliers of timber for mine shaft supports, construction, and fuel, this population increased to the point where it seceded from the Sacred Heart Church, secured permission from Alexander Agassiz to use the site on which Saint Patrick Hall stood, and then acquired the hall for thirty-five hundred dollars. Bishop Vertin consecrated the hall Saint Louis Church.

By 1899, when the Saint Louis congregation had grown from 104 to 375 families, members raised sufficient funds for a new church. With the appointment of J. R. Boissonault as priest the previous year, a building committee of Joseph Desjardins, Alex Ethier, Joseph Chatet, and Joseph Ouelette was formed, and in 1900 the bishop gave permission for the congregation to build a church.[44] However, he forbade them to incur a debt greater than ten thousand dollars. The building committee selected an architect and announced its intent to build a new church.

The *Copper Country Evening News* for 7 March 1899 described the proposed church and discussed the improvement the structure would represent to the congregation and the community:

> The location on the corner of Scott and Fifth streets opposite the Union Building, is a fine location and the new church . . . will be an ornament to the locality. The plans call for a building of solid sandstone and the style of the architecture will be Gothic. . . .
>
> The building will be 47 by 130 feet in extreme ground dimensions and a large tower, calculated to hold a chime of bells, will rise from its southwest [*sic*–northeast] corner to a height of 130 feet while the rear or north [*sic*-south] corner of the building will be marked by small ornamental turrets.
>
> The basement will contain a chapel and three large class rooms, besides the heating plant, while the main floor of the church will seat 400 people, the octagonal sanctuary, flanked by two side altars, being lighted by windows above similar to those in the sanctuary of the cathedral at Marquette. Three large entrances will give admission to the church, which will adjoin the present priest's residence, the lot being 100 by 150 feet in size.

Charlton, Gilbert and Demar were the architects, and their design seemed to suit exactly the needs of the congregation. By 1900 the firm had offices in Hancock, Sault Sainte Marie, and Marquette, Michigan, and in Milwaukee, Wisconsin.

Prendergast and Clarkson of Chicago, builders of the Copper Range Passenger Station and General Offices in Houghton and several large commercial blocks in Hancock, began construction in May 1900. Using a new labor-saving steam-hoisting device never before seen in the Copper Country, workers elevated the huge stone blocks, three and four at a time, and boxes of mortar from the ground and swung them to any position on the wall. A large force of masons quickly laid stone in the walls.[45]

Randomly coursed, rock-faced square and rectangular blocks of stone diminish in size as they rise in the exterior walls of the church. Smooth-cut and chiseled stone trims openings. Wall buttresses are applied between the windows to the long side walls of the rectangular building. An attached corner tower on the northeast soars 130 feet in three stages to an open belfry surmounted by an octagonal spire. The corners of the first two stages are supported by pier buttresses, and engaged conical turrets mark four sides of the octagonal third stage. A triple pointed arched entrance on the north porch gives access to three small vestibules and to the four-hundred-seat nave.

The organ and choir loft approached by a stair at the northeast corner of the church is above the vestibule and extends twenty feet into the nave. A circular baptistery was at that time at the northwest corner. The octagonal sanctuary is flanked by side altars. The basement held a chapel, classrooms, and the heating plant.

Stained glass windows in dark purplish blue, pale reddish brown and green with center images depicting saints fill the window openings on all walls of the church. They are memorials to Valliéres, Laberge, Desjardins, Ethier, Crépeau, Primeau, Ouellette, and other French-Canadian individuals and families.

Construction of Sainte Anne's Church was complete one year after it began, and a throng of worshipers, church officials, and visitors from the Keweenaw Peninsula dedicated the church on Sunday, 16 June 1901. The *Copper Country Evening News* for 17 June 1901 said the dedication was unequaled in the history of the Copper Country churches. Over five hundred people attended the three-hour ceremony officiated by clergy from as far away as Montreal and Iowa. Some eight hundred members of Catholic societies—Croatian, Polish, Italian, French Saint Jean Baptiste societies, Saint Anthony Court of Catholic Foresters, and the Ancient Order of Foresters of America—participated in the services. Filling Fifth Street as far back as Elm Street in the largest parade ever seen in Calumet, the members of the societies marched in full regalia into the church by the main entrance and out by the side entrance, each contributing money as he passed through. Invited guests and the public were admitted until the nave, gallery, and standing room were filled with more than eleven hundred people. Following the dedication, the Ladies of Sainte Anne's Church served dinner, also an occasion to raise funds.

Sainte Anne's Church is one of the many great ethnic sandstone churches that gave Calumet the name, "the city of churches." Another is the Slovenians' and Austrians' Saint Joseph's Austrian Church (Saint Paul the Apostle Church) of 1903–8, the double-spired, Richardsonian Portage Entry red sandstone extravaganza at 301 Eighth Street probably designed by Charles K. Shand, which rises authoritatively over the village like a cathedral of medieval Europe. Still another is Saint Mary's Italian Church, a small central towered Gothic structure built in 1896–97 on the south side of Portland Street between Ninth and Tenth Streets. Sainte Anne's Church and Saint Joseph's Austrian (Slovenian) Church are among the last solid sandstone buildings ever constructed in Calumet and speak of the importance of religion and ethnicity to the French-Canadian, Austrian, and Slovenian Catholics of Calumet. Now deconsecrated, Sainte Anne's Church is undergoing rehabilitation and adaptive reuse as the Keweenaw Heritage Center.

Laurium

Laurium is a residential community originally populated by miners and business people from Calumet. Platted on a grid and incorporated as a village in 1889, Laurium grew to about three hundred acres in 1900. The village has a commercial district, Portage Entry sandstone village hall, and many fine houses, as well as clusters of mining company housing.

Paul P. and Anna L. Roehm House, Laurium

In 1895–96, when Laurium achieved substantial growth as a residential community for managers of the Calumet and Hecla Mining Company, as well as bankers, professionals, and merchants in Calumet, Paul P. Roehm (1857–1925) built a large sandstone house for himself and his wife Anna (1856–1945) at 101 Willow Avenue (the northwest corner of First Street) in Laurium. Roehm, the region's preeminent stone mason and supplier, may have designed and built the house himself.

The rock-faced, Portage Entry red sandstone house achieves a vernacular character from its simplification of the Richardsonian forms. The unpretentious tower, capped by a sloping conical roof that rises out of the massive picturesque composition, demonstrates this vernacular adaptation. The heaviness also derives, in part, from the stone building material, from the muscular random masonry walls, and from the roughly hewn and tapered stone piers of the porch.

The house is thoughtfully adapted to the climate. The entry hall, dining room, drawing room, and library pivot around a central chimney, with fireplaces placed on a diagonal in the core of the building in a manner not unlike seventeenth-century British Colonial works in New England or the early Frank Lloyd Wright work, the Winslow house (1893). Ample fireplaces provide warmth, and sliding doors permit the closing of rooms to contain heat and reduce drafts. South-facing windows in the tower admit light into a reading alcove of the library, and the early morning sunlight enters a sun room at the rear. Beyond is a sitting room and kitchen.

Roehm house, 1898, Demetrius Frederick Charlton. From W. E. Steckbauer and Albert Quade, A Souvenir in Photogravure of the Upper Peninsula of Michigan *(Brooklyn, N.Y.: Albertype Co., 1900).*

Roehm's stone yard became the primary outlet locally for stone from the Portage Entry quarries, and Roehm transacted business almost daily with J. W. Wyckoff, superintendent of the Portage Entry Quarries Company, and his customers and clients. Moreover, Roehm built many stone commercial blocks, banks, town halls, and churches. Among his major projects were the Kinsman Block, the Red Jacket Town Hall and Opera House, Saint Joseph's Austrian Church in Calumet, and the Sacred Heart Church in Laurium.

As it neared completion, Roehm's palatial house was regarded as a major investment worthy of the community. The Roehm house and the shingle-style Johnson and Anna Lichty Vivian House—a building that combines red Portage Entry sandstone at the first level with shingles at the upper level and is located at the northeast corner of Pewabic and Third Streets—together illustrated W. E. Steckbauer's and Albert Quade's *A*

Souvenir in Photogravure of the Upper Peninsula of Michigan (Brooklyn, N.Y.: Albertype Co., 1900). Thus, the stature of Laurium was reflected in the stature of its marvelous red sandstone houses.[46]

PAINESDALE

Painesdale is an early twentieth-century mining community less than ten miles southwest of Houghton that was planned, financed, and managed by East Coast developers and inhabited by immigrant miners and their families. It is named after William A. Paine of Boston, founder of the Painc, Webber and Company brokerage firm, who was a chief investor in the Copper Range Company and its president for thirty years. The Copper Range Company promoted and developed a section of the mineral range southwest of Portage Lake that contains the Atlantic, Baltic, and Isle Royale lodes. The Champion Copper Mine located near Painesdale was an important producer from 1899 to 1916, and, under the direction of its parent corporation, the Copper Range Company, it continued to produce intermittently until 1967. Rows of stock-designed miners' houses with sharp gable roofs and shingled or clapboarded siding form a community image of efficiency and homogeneity in Painesdale, even today.

Originally, the Champion Copper Company's holdings consisted of approximately forty buildings and structures, including four shaft houses and four hoist houses, a railroad depot, an office, and several boiler houses. They include handsome mine rock and sandstone masonry buildings like the E Shaft hoist house and the nearby machine shop, both rectangular buildings with gabled roofs and cut coursed sandstone walls (1902).

North and west of the mine are located four residential enclaves of worker housing, known as locations, and a separate district containing officials' houses. B Location, C Location, E Location, and E Addition, or Seeberville, developed from north to south alongside operations following the course of the Champion copper lode.

East of C Location, the Copper Range Company landscaped a public square and around it built the Sarah Sargent Paine Public Library, the Albert Paine Methodist Church, and the Jeffers High School. The library and an elementary school no longer stand, but the towered Neo-Gothic church (1907) clad in shingles and clapboards still marks the northwest corner of the green. The huge Neo-Tudor red sandstone high school (1909–10, 1935) on the east side of the square with a distant view of Keweenaw Bay to the east is the most elaborate structure in Painesdale. Its rock-faced stone walls, coursed evenly on the raised basement and randomly above, are articulated with smooth-cut window surrounds and quoins. The buttressed projecting entrance pavilion has a richly carved Tudor arch.

Jeffers High School (Painesdale High School), 1909–10, 1935, Alexander C. Eschweiler?, Painesdale, photograph c. 1985. (Courtesy Balthazar Korab)

Sarah Sargent Paine Memorial Library, Painesdale

William A. Paine built the Sarah Sargent Paine Memorial Library as a memorial to his mother for the benefit of the employees.[47] The Neo-Tudor library was constructed of fine-quality red sandstone to serve as an educational and social center for the workers of the Champion, Trimountain, and Baltic Mining Companies. The Copper Range Consolidated Mining Company wanted to maintain a contented work force. Prendergast and Clarkson of Chicago constructed the library between 1902 and 1903.

Paine selected Alexander C. Eschweiler (1865–1940), an architect thoroughly familiar with the Copper Country, to draft plans for the library. Born in Boston, the son of a mining engineer, Eschweiler had spent his early childhood and youth in the Copper Country. In 1882 he moved with his family to Milwaukee where he attended Marquette College. Eschweiler studied architecture at Cornell University and worked as a draftsman in several offices in Milwaukee. In 1916 and in 1921 he was joined in practice by his sons A. C. Eschweiler, Jr. and Theodore L. Eschweiler, respectively. He established the firm of Eschweiler and Eschweiler in association with these two sons and a third son, C. F. Eschweiler. During his career Eschweiler designed numerous churches, schools, offices, industrial buildings, hospitals, and houses.[48]

The library occupied the main floor of the building with a general reading room, a stack room for the books, a smoking room, and a reading room for children. The rooms were provided with chess and checker tables and with other games. The second floor held the auditorium, which was equipped with a stage. Public baths were installed in the basement.

A few years earlier, in 1898, the Calumet and Hecla Mining Company had built a library and baths for its workers in Calumet. Italian masons meticulously laid cut dark gray basalt and gray and rose granite in its exterior walls, and the whole was trimmed with orangish red brick. The architect was George Russel Shaw of Shaw and Hunnewell of Boston. When the Calumet Library opened, the *Copper Country Evening News* commented that "no mining company in the world treats its employees better than the Calumet and Hecla. It has just completed a large library building which will be provided with a fine collection of books, many new ones having been added to the old list, which is free to the employees. There are baths in the building and many comforts for the workingmen to enjoy." Indeed, President Agassiz instructed the library committee on the building regulations that were written to encourage the miners and their families to use the library freely.[49]

Dedication ceremonies for the Sarah Sargent Paine Memorial Library took place in the auditorium of the building on 14 November 1903 with F. A. Jeffers, superintendent of the Adams Township schools, addressing the audience of friends and members of the Paine family and members of the community.

The responsibility to hold, control, and manage the library was vested jointly among Paine; Lucius L. Hubbard, general manager of the Champion Copper Company; and Frederick W. Denton, general manager of the Trimountain Mining Company, also of Houghton County. The Atlantic, Baltic, and Trimountain mines were obligated to contribute financial support to the library. At regular intervals, cash donations would be received and, in turn, library books would be carried in large wooden trunks between mine offices for the employees and their families.

The labor unrest in the Copper Country in the 1890s and early 1900s culminated in a general strike called by the Western Federation of Mines on 23 July 1913 of all the members employed in the Copper Country and resulted in massive unemployment for over fourteen thousand men,

Calumet and Hecla Library, 1898, Shaw and Hunnewell, Calumet, Keweenaw National Historical Park, photograph date unknown. Boston architects designed the library and the nearby general office building, and Italian-American stonemasons laid granite and poor or waste mine rock in the exterior walls. (Courtesy Michigan Technological University Archives)

including two thousand at Painesdale. Among the grievances were the poor living conditions. Through the erection of a substantial stone library for the miners of this region, William A. Paine and the Copper Range Company intended to alleviate and avoid dissatisfaction among workers and to state symbolically the benevolent and paternalistic attitude of the mining companies. With a cultivated high style library in native red sandstone, management tried to convey the worthiness of the community for such an investment and the capability of the company in looking after its welfare. The library was destroyed in 1964.

Summary

Like Marquette, no professionally trained architect arrived in the remote Copper Country until the 1890s. Clients called upon architects from Chicago, Milwaukee, and Detroit to design the region's major public buildings. In 1889 John Scott of Detroit designed the Michigan Mining School and the First National Bank of Houghton; in 1899 Henry L. Ottenheimer of Chicago designed the Shelden-Dee Block and a group of commercial buildings in Houghton; and in 1902 Arthur C. Eschweiler created the Sarah Sargent Paine Memorial Library at Painesdale. All three architects were well-trained professionals who had developed reputations as first-rate architects. The big-city architects were selected by sophisticated easterners, men and women with taste, vision, money, and power, who were land speculators, stockbrokers, mining company engineers and managers, scientists, investors, and lawyers. Although the architects had offices in eastern and midwestern cities, they developed ties with the Upper Peninsula. Scott and Ottenheimer designed a number of public buildings there and opened offices, the former in Marquette, the latter in Houghton. Eschweiler, who had grown up in the Copper Country, designed the Painesdale High School. These architects and their clients, people who had profited from the geological resources of the area, realized the appropriateness of using the native sandstone building material and used it in their cultivated designs. These men and their works guided and influenced the taste of the others who had little formal training.

Local governments, businessmen, and religious groups called on Demetrius Frederick Charlton, the first trained architect in the Upper Peninsula, or on itinerant architects, such as Charles K. Shand, who followed the development westward across the country, to design their city halls, courthouses, churches, banks, and commercial blocks. In 1898 Edward Demar of Charlton, Gilbert and Demar planned the Hancock Town Hall and Fire Hall. One year later Charlton designed Sainte Anne's Roman Catholic Church for a French-Canadian parish at Red Jacket. For a client seeking the services of a well-trained local architect, Charlton was the man to turn to. The trip from Marquette to Hancock, Calumet, Lake Linden, and Laurium was an easy one by train, and Charlton could visit the building sites. Moreover, Charlton actively and successfully competed for the important commissions in the region. Between 1898 and 1899, Charles K. Shand designed the Red Jacket Fire Station, an addition and alterations to the Red Jacket Town Hall and Opera House, and in 1902–4 Saint Joseph's Austrian Church.

Men with little professional training but experienced as builders built most of the Copper Country's architecture. In the 1890s C. Archibald Pearce designed the Finnish Lutheran College and Seminary, modeled to some extent after the Michigan Mining School by then standing in Houghton, but he treated it more directly in an unaffected way. In 1897 Paul Roehm, a skilled stone mason, designed and built his own house in Laurium.

Throughout the region, there are industrial structures of Jacobsville sandstone, some made of squared blocks of stone, but most composed of rubble stone combined with poor mine rock. Mining buildings of all kinds—rock houses, machine shops, blacksmith shops, power houses, stamping mills, foundries—stand vacant and in ruins today, clusters of red and black rock. They were constructed by the engineers and workmen of the mining companies themselves. The company office buildings, libraries, bathhouses, and schools, however, are often of carefully extracted and finished blocks of sandstone after the designs of trained architects and engineers. The Quincy Mine Company Office, designed by Robert C. Walsh of Morristown, New Jersey, where the mining company's home office was located, and built in 1895–97, matches in skill the designs of the region's vigorous and vital churches, banks, and commercial blocks. The vernacular tradition persists in the sandstone architecture of the Copper Country to an even greater extent than that of Marquette.

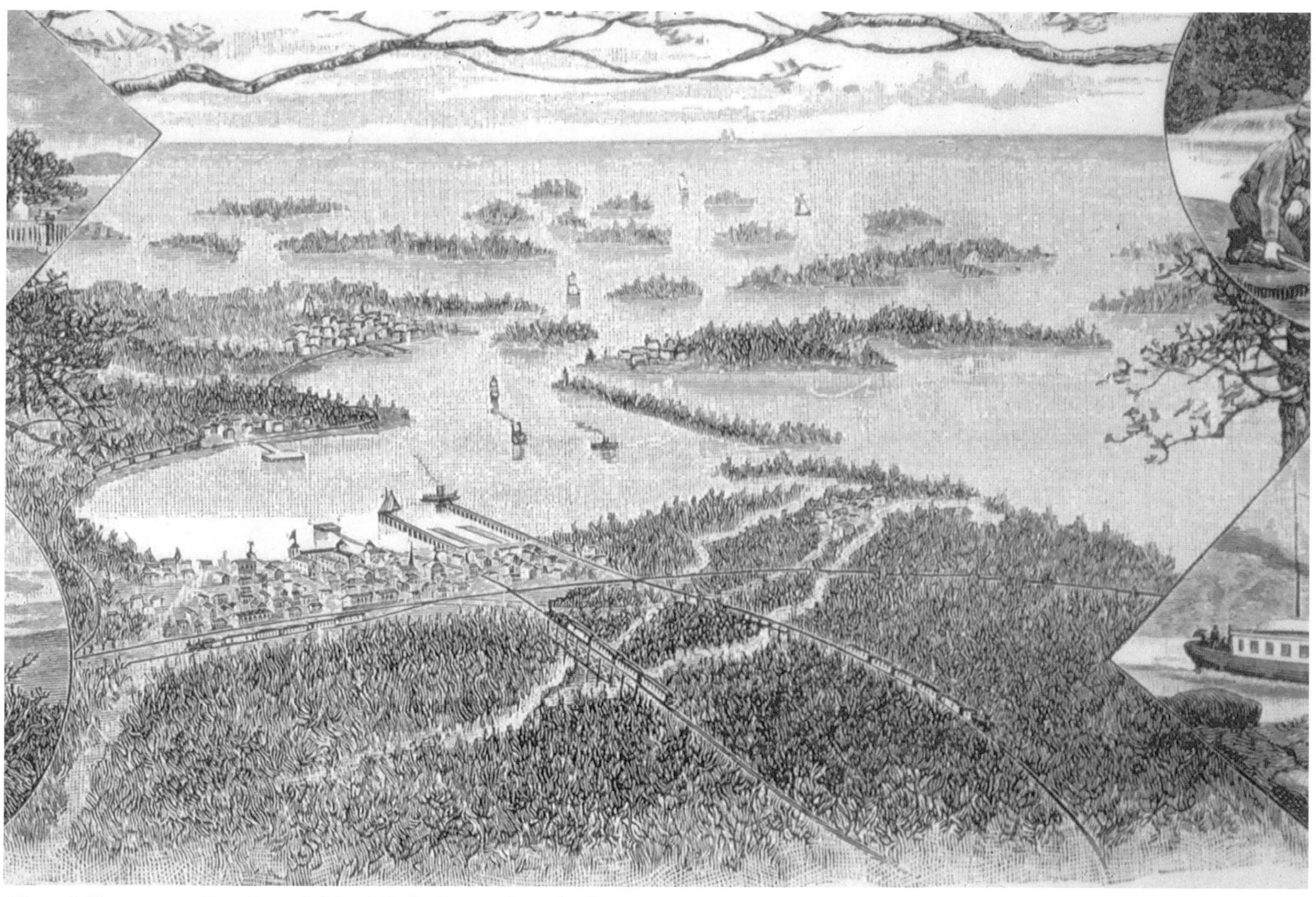

View of Chequamegon Bay. From Ashland Daily Press, *Annual Edition (1893).*

4

THE SANDSTONE ARCHITECTURE OF THE CHEQUAMEGON BAY AREA, SUPERIOR, AND DULUTH

INTRODUCTION

The plain of the Lake Superior Lowland extends along the south shore of Lake Superior from Ashland, Bayfield, and Douglas Counties in the northwest corner of Wisconsin to the head of the lake at Duluth, Minnesota, and to the rugged hills beyond. Here the sand and clay soil is better suited for grazing and growing hay than for general agriculture and grain production. Ravines and hills relieve the plain. An escarpment extends from the southwest at the Wisconsin-Minnesota border northeast to the Apostle Islands, marking the boundary between the Lake Superior Lowland and the Northern Highland. The land curves along the shore of Chequamegon Bay rising to rocky hills that are the site of Washburn and Bayfield. The Apostle Islands shelter the bay from the fierce storms and turbulent waters of Lake Superior. Where streams and rivers flow into Lake Superior, little fishing villages like Cornucopia and Port Wing became the trade centers for neighboring farmers. Ranges of rolling hills and valleys are timbered with pine and hemlock forests and hardwoods.

Major settlements on Lake Superior west of Keweenaw Bay cluster around two large natural harbors: Chequamegon Bay, 350 miles west of Sault Sainte Marie; and Duluth-Superior Harbor at the head of the lake. The harbors of Ashland, Washburn, and Bayfield are situated on Chequamego Bay. Chequamegon is a French translation of the Chippewa word Shaugawaumekong, meaning "a long narrow strip of land running into a body of water."

Pierre Espirt Radisson and Médard Chouart, Sieur de Groseilliers, traveled along the south shore of Lake Superior, landing at Chequamegon Bay in 1659 and returned to Montreal and Quebec with reports of the wealth of furs found here. This information led to the establishment of the Hudson Bay Company. French explorers, missionaries, and fur traders quickly followed Radisson and Chouart into the area. Furs formed the basis of the economy until 1834, when John Jacob Astor of the American Fur Company sought richer hunting grounds to the West.

Stella Grove, Presque Isle (Stockton Island), Apostle Islands, photograph 1898. (Courtesy Library of Congress, Detroit Publishing Company Photographic Collection)

Bayfield

Bayfield lies twenty-two miles north of Ashland on a forested rocky hillside down which streams run into Lake Superior. The settlement was platted in 1856 after the Bayfield Land Company was formed to finance the construction of a railroad between the wheat fields of the West and this Lake Superior port; it was named for Lieutenant Henry Bayfield of the British Navy, surveyor of the Great Lakes who first charted Lake Superior in 1823–25. Today Bayfield's small permanent population expands to many more during the summer resort season.

Henry M. Rice (1816–94), a member of the Minnesota territorial legislature, eagerly sought to connect Minnesota with a port on Lake Superior for economic reasons. Rice obtained a federal land grant to build a railroad from Saint Paul, Minnesota, to Chequamegon Bay and convinced investors in Washington, D. C., to form the Bayfield Land Company.[1] Visions of a settlement that would rival Chicago in commerce and in grain shipping attracted speculators to Bayfield. However, the panic of 1857 ruined the Bayfield Land Company, and the Chicago, Saint Paul, Minneapolis and Omaha Railroad did not arrive until 1883. In the meantime, other investors exploited and developed the area's timber, fish, and sandstone.

Robinson D. Pike opened a sawmill in 1869 that produced boards, shingles, and barrels; the Boutin family established in 1870 a commercial fishing business that eventually employed 150; and the Booth Fisheries opened in 1880, and by 1890 the business employed 150. The scenic natural beauty and cool breezes attracted summer resorters to Bayfield. The first sandstone quarry was opened in 1868, and by the 1880s seven sandstone quarries operated in the area. In fact, one writer noted that "large quantities of some of the finest building material in the West are obtained from the red sandstone quarry in the harbor."[2]

Bayfield County Courthouse (Apostle Islands National Lakeshore Headquarters), Bayfield

The first sandstone building erected in northern Wisconsin was the Bayfield County Courthouse. The courthouse stands in a public square on the highest plateau in the center of the village of Bayfield overlooking the Bayfield harbor and the Apostle Islands.

On 14 February 1883 fire destroyed the first Bayfield County Courthouse, a wooden structure built by B. F. Bicksler, a local carpenter, ten years earlier.[3] Within days, the people of Bayfield rallied for a new courthouse. The Bayfield Businessmen's Association and the Bayfield County Board of Supervisors quickly met to plan a new building. Although officials and citizens disagreed over the style, cost, and location of the structure, most felt that work should begin at once. In the ensuing weeks, they developed a scheme for a prominent fireproof public building within the county's budget that would solidify the stature of Bayfield and promote one of her natural resources. From the outset, it was clear that a courthouse executed in native Bayfield brown sandstone would symbolize the aspirations of the people of Bayfield. Indeed, the Bayfield County Courthouse would be the first sandstone building erected in northern Wisconsin.

The Bayfield Businessmen's Association adopted resolutions concerning the urgency, cost, and location of the proposed courthouse and submitted them to the county board. The association urged the county board to build immediately a courthouse and jail. It felt that no more that twenty thousand dollars should be spent improving the site and in building the structures, and it thought the courthouse and jail should be placed on the public square.[4] Inherent in the resolution was the commitment to build in native brown sandstone.

That spring, several members of the businessmen's association ran for seats on the county board. Their support of a new courthouse swept them into office. The new board included Frank Boutin, Sr., a fish dealer and lumberman; William Knight, Indian agent, banker, investor, and horticulturist; Fred Fischer, a timber speculator and merchant; and Robinson D. Pike, a lumberman, sawmill operator, real estate investor, and owner of a nearby yet-to-be-developed brownstone quarry. Pike, Bayfield's leading businessman and citizen, served as chairman of the newly elected board.[5] These men promoted the growth and development of Bayfield just as a chamber of commerce would today, and they assumed this same fervor in their involvement in the courthouse project.[6]

The Bayfield businessmen struggled to secure the investment of outside capital for the development of resources in their town in the West, but prospective eastern and urban investors needed assurances their investments would be sound. The businessmen would demonstrate the solidity of Bayfield through the construction of a prominent and substantial courthouse of native Bayfield sandstone. Constructing a public building of fireproof stone from their own sandstone quarries would advertise their newly developing sandstone industry.

The day after the fire the board decided to erect a temporary building on the site of the ruined courthouse. The board then began listing its functional, spatial, and other requirements that would guide an architect in developing a design for the new courthouse. It invited and studied proposals for a new courthouse from several architects and conferred with others engaged in erecting public buildings. From Edward Townsend Mix, the fully trained and experienced Milwaukee specialist in public building design, it solicited a sketch. Not fully satisfied with Mix's work, the board corresponded with the Houston County, Minnesota, commissioners to inquire about the courthouse that county was constructing.[7]

Several plans for the new courthouse were exhibited in the village the last week of March, and the *Bayfield County Press* remarked on those procured by a Mr. Patterson that showed "an elegant two story, with basement, brown stone court house, which the architect says can be erected for $20,000."[8]

One week later on 7 April, the newspaper described the plans of a Saint Paul architect:

> The county board is in receipt of a plan for a brown stone courthouse, from a St. Paul architect, which can be erected for $20,000. It contemplates a large two story building, with basement and fireproof vaults, and would be an ornament to any town in the northwest. We believe the board has not decided to accept any one of the several plans received all of which are good but propose[s] to make haste slowly, thereby insuring

no mistakes. We venture the prophecy, however, that Bayfield county will have a stone court house and at a cost that will not prove burdensome to any tax payer in the county.[9]

Finally, on 19 April 1883, presumably after reviewing his draft plans, the board invited John Nader of Madison, an engineer and surveyor, to travel to Bayfield to discuss the plans and construction of the building.[10] The board selected Nader as the courthouse architect.

In selecting John Nader, Pike, Boutin, and Knight kept the promise made during the election campaign—to erect the new building of native brownstone in the public square and equip it with all modern improvements. Locating the new courthouse in a more commanding position was a principal consideration of the board because its former site on the flat had shown off the courthouse to poor advantage. The board considered two sites: block ninety-five on the corner of Rittenhouse Avenue and Sixth Street; and the public park, one-square block higher up the hill donated to the village by Henry M. Rice. With Nader's encouragement, it chose the public square overlooking the Bayfield harbor and the Apostle Islands.

While conferring with the board on the plans for the building, Nader inspected the grounds. The *Bayfield County Press* for 5 May 1883 stated: "Mr. Nader reported the public square by all odds the finest location in the village for the court house and the board has decided to locate it thereon."

During the visit, Nader assured the board that for twenty thousand dollars the county could have a beautiful fireproof building. The local newspaper elaborated further, stating that the courthouse will be a timeless ornament to Bayfield and "an advertisement for one of our chief resources—our brownstone, of which he [Nader] is outspoken in pronouncing the best building stone in the country, and in fact about the only really fire proof stone known, having stood the test of the Chicago fire where all others failed."[11]

Nader completed his plans within two weeks. He interpreted the board's wishes by carefully calculating and designing a plain and simple restrained Italianate structure, one that might be expected from a trained and experienced engineer and surveyor.

Before coming to Wisconsin from the New York City area in 1868, Nader (1838–1919) designed and built marine fortification structures in the East. He served as assistant United States engineer in charge of the Wisconsin River Improvement, headquartered in Portage in the early 1870s, and as city surveyor for Madison from 1876 to 1880, and again from 1884 to 1887. Nader first listed himself as an architect in 1883, the year he designed the Bayfield County Courthouse. In 1887, after designing his best-known work, Saint Patrick's Church, constructed at Madison in 1888, he left Wisconsin for Virginia and New York City.[12]

The first elaboration of Nader's plans for the courthouse appeared in the *Madison Democrat* for 5 June 1883. The article stressed again the employment of the native sandstone as a means of promoting the native resource and industry:

> The county court house, a modest frame building was consumed by fire last winter, and the commissioners decided to rebuild in stone and in this respect the county is unanimous and while supplying a necessity, to develop one of the resources of the country. The superior sandstone is located here in abundance, and will provide the north with building material for several years.
>
> Architect Nader, of this city, has provided the plans for the new building which will be erected during the present season. The masonry of the building will be entirely of brown sandstone obtained on the main land. Private contributions will place a bust of Capt. Bayfield, the first explorer, of the region, over the main entrance, to be done in the native sandstone. A clock will also be contributed, to be placed in the tower with five and one-half foot dials, 63 feet above the main floor.
>
> The location of the building is chosen so that it will present a fine appearance from the bay south and east of Bayfield. A spring was discovered of sufficient head to provide the highest point of the building and the architect has made the most of it for all purposes.[13]

The board examined, discussed, and approved the plans, all on 21 May 1883. Then the county clerk advertised for bids to build "a Stone Court House at Bayfield" in the (Saint Paul) *Daily Press,* the *Daily Milwaukee Sentinel,* and the *Bayfield County Press.*[14]

On 22 June 1883 the board awarded the construction contract to Cook and Hyde of Milwaukee, operators of the sandstone quarry on Basswood Island. Their bid of twenty-nine thousand dollars included the full erection of the structure but did not include the heating apparatus, vault fronts, jail doors, and burglar-proof chests.

On 20 June 1883 the *Bayfield County Press* recapitulated the events leading to the erection of the new courthouse and presented a plan of the building as well. It noted that the rough-cut, native brown sandstone for the courthouse was taken from Pike's quarry at Van Tassell's Point, and the cut stone from Cook and Hyde's quarry at Basswood Island, a fully established quarry with a saw mill.[15] Both quarries were at water's edge, and their products easily transported by ship to Bayfield.

Construction on the courthouse began in the summer of 1883. Workmen, under the supervision of a man named Gleason, completed the

Bayfield County Courthouse (Apostle Islands National Lakeshore Headquarters), 1883–84, John Nader, Bayfield, photograph c. 1885. This historic photograph shows the original clock and bell tower. (Courtesy Apostle Islands National Lakeshore)

basement by the first of September. They laid the stone in the walls, put on the roof, and hung the windows and doors that fall, thereby enclosing the structure before work was suspended for the winter. Work resumed in mid-April and was completed by July 1884. The project cost approximately thirty-one thousand dollars.

The Bayfield County Courthouse is a symmetrical, two-story Italianate public building on a raised foundation. It measures some seventy feet in width by fifty feet in depth. Variegated Bayfield sandstone rough-cut and evenly coursed rises in the exterior walls. The exterior is trimmed with smooth-cut and tool-finished corner quoins and pieces of the same stone. Pedimented central entry pavilions project slightly from all four facades with wooden steps ascending to entries in all but the rear. The facades are divided into five bays on the front and rear and three on the sides. A hip roof with deck, originally topped with a short tower containing a four-dial clock and bell, crowns the structure. The arched main (south) entry is surmounted by a fanlight and stone cornice and flanked by stone pilasters. Windows, segmentally arched on the ground and first stories, and round arched on the second story, increase in height as the building rises from the basement to the second floor. The word "Bayfield" carved in bold letters in the center of the front pediment ties the building to the county.

Originally, offices for the register, treasurer, clerk, judge, sheriff, and district attorney flanked a central corridor that ran the full depth of the building on the first floor; the courtroom, and rooms for the judge, jury, and witnesses occupied the second floor; and the jail and a boiler room took up the basement. The building was heated with steam and supplied with water from a spring in the hill north of the building.

Bayfield County abandoned this courthouse in 1892 when it moved the seat of county government to Washburn. Subsequently, the building has served as a town hall, school, community center, and quarters for prisoners of war during World War Two. In 1978 the Bayfield Heritage Association acquired the building and restored the exterior and renovated the interior with funds from the Economic Development Administration and the Upper Great Lakes Regional Commission. The building is currently leased to the National Park Service and used as the headquarters for the Apostle Islands National Lakeshore.

By 1883 Ashland and Washburn emerged as rival deep-water ports on Lake Superior and challenged Bayfield's preeminence. That year the arrival of the Chicago, Saint Paul, Minneapolis and Omaha Railroad at Washburn established that village as a principal port on Chequamegon Bay. Eventually, in 1892, Washburn was designated as the county seat.

Through the efforts of William F. Dalrymple, brother of Oliver Dalrymple, Bayfield attempted to regain its prestige with plans to build a railroad north along the shore of Lake Superior to the Dalrymple bonanza wheat farms in the Red River area of North Dakota and Minnesota. The death of William F. Dalrymple and the panic of 1893 ended Bayfield's attempt to recapture its past glory. Only three and nine-tenths miles of track were laid.

The comments of visitors to Bayfield demonstrate the success of the courthouse in expressing the values of Bayfield residents. Investors from Saint Paul and New York City, assessing Bayfield's prospects for commerce, business, and tourism in December 1883, thought the site, material, and architectural appearance of the courthouse reflected "taste, economy, and practical sense" in the construction of public buildings.[16] The editor of the *Lumberman* visited Bayfield in February 1885 and called the new courthouse the most noticeable improvement to the town since his last visit: "I must say it is the finest one I have seen in the Northwest and does infinite credit to the town and county. . . . It was built in the highest portion of the town of the Bayfield brown sandstone, and from the tower overlooks the whole bay and all of the islands and lake as far as the eye can reach, as well as the opposite shore, with Ashland in plain view."[17]

On the twenty-ninth anniversary of the settlement of Bayfield, the *Bayfield County Press* for 28 March 1885 reflected on the community's accomplishments. It thought the use of native brownstone in the Bayfield County Courthouse of particular importance to Bayfield's identity. The *Press* proclaimed proudly, "Now we boast the most stately structure in Northern Wisconsin, Bayfield County court house. . . . rendered still more a source of pride from the fact that its component elements are native to our county. This leads us also to call to mind that within a very few years our brown stone has come into favorable notice and bids fair to rival all of the best known kinds of stone for building purposes."

For more than twenty-five years, Bayfield people continued to build with Bayfield sandstone from quarries on the Apostle Islands and the mainland. The Bayfield School Board constructed Lincoln School of solid brownstone in 1894. Robinson D. Pike saw to it that most projects he was associated with employed local sandstone, and he even occasionally donated the material from his quarry for others. Thus, the parishioners of Holy Family Catholic Church employed Bayfield brownstone donated by the Pike quarry in the construction of their church, a ten-year project that concluded in 1898. And the Wisconsin State Fish Commission used sandstone and wood shingles in the State Fish Hatcheries. The fish hatcheries were built in 1895 on land donated by Pike at Salmo, two miles southwest of Bayfield. Bayfield builders routinely laid foundations of brownstone and trimmed the village's most important structures with the material. In this way, the library board built the Carnegie Library of yellowish orange brick trimmed with Bayfield sandstone to the designs of Henry E. Wildhagen in 1903.

First National Bank, Bayfield

Motivated in part by an intense rivalry with neighboring Washburn and a desire to reverse the effect of the decline of the region's lumbering and quarrying industries, a group of Bayfield investors established a bank and built a small brownstone bank building in Bayfield just after the turn of the century, in 1905.

Early in 1904, M. A. Sprague, F. T. Yates—publisher of the *Washburn News and Itemizer*—and O. A. Lamreaux from nearby Washburn filed articles of incorporation for the Bayfield State Bank with the Wisconsin State Commission on Banking. To the community this action and the news that the Omaha Railroad was considering building an extension to Bayfield seemed to forecast prosperity. The plans of Washburn subscribers to form a state bank in Bayfield inspired Bayfield capitalists to establish a national rather than a state bank. With a capital stock of twenty-five thousand dollars, the Bayfield men applied to the controller of currency to establish the First National Bank of Bayfield. The new national bank would replace the Lumbermen's Exchange. Eventually the competing groups would join forces to establish one bank—a national bank.

One year after the First National Bank of Bayfield was established, it had outgrown its quarters, and the directors and officers decided to build a fireproof brownstone bank building in the midst of Bayfield's approximately four-block commercial district on a site on the northwest corner of North Second Street and Rittenhouse Avenue. The Washburn promoters of the state bank had acquired this site, known as the Pike property, earlier. The local newspaper announced that the plans for the bank building were under preparation and that the structure would be as complete in all details for its size and as fireproof as any in the country.[18] The new bank seemed to demonstrate the power of cooperation.

Robinson D. Pike, although apparently not involved in the bank venture, incorporated an office building for himself into the plans for the bank. Nothing is known of the architect or designer for the bank or office. But on 12 May 1905, as Pike removed an old house on the corner of North Second Street and Rittenhouse Avenue adjoining the bank property to clear the site for his brownstone office building, the *Bayfield County Press* explained that the office would correspond to the design of the new bank building and would add much to its appearance. The bank and the office took shape as a single unit.

Brownstone for the First National Bank probably came from Pike's own quarry on the bay at Van Tassell's Point, one mile south of Bayfield. The Pike quarry contained large quantities of excellent quality Bayfield brownstone. Operations at the quarry began in 1883. By 1910, five years after the bank and office building opened, there no longer existed a market for colored sandstone, and the brownstone industry fell into decline.

Plans for the First National Bank were completed by the end of April 1905, and construction began soon after. Opening bids for the erection of the First National Bank were made on 2 May 1905, and the bank directors let the contract to talented Ashland stonecutter Archie Donald.[19] The bank was completed in four months, Pike's office in seven.

The exterior walls of the modest one-story Beaux-Arts Classical bank and office building are constructed of large blocks of clear brownstone ashlar laid evenly. Piercing the walls are large single-paned windows with cross-paned transoms, arranged in a pair and a group of three on the Rittenhouse Avenue facade and arranged singly on the North Second Street facade. Three steps lead from the street to the main corner entrance of the bank. A side entrance over which is carved the name "R. D. Pike" gives access to the office. Through its strictly formal design executed in sturdy brownstone, the little building achieves a simple monumentality.

At the time the people of Bayfield and the surrounding region sought ways to revitalize their sagging economy. Although commercial fishermen continued to ship tons of herring, trout, and whitefish from packing houses at Bayfield to all parts of the country, by 1905 the forests of the region were depleted. Businessmen, real estate investors, and farmers intended to devise ways of promoting the region as a place well suited for the farmer, stock raiser, fruit grower, and dairyman.

With the exquisitely crafted and dignified Bayfield brownstone bank and office building, local businessmen valiantly proclaimed that Bayfield had passed the early period of settlement and growth and was capable of taking charge of its future.

Washburn

In 1883, more than twenty years before the erection of the First National Bank at Bayfield, the Chicago, Saint Paul, Minneapolis and Omaha Railway Company extended a line north along the western shore of Chequamegon Bay eleven miles north of Ashland and purchased the site of Washburn. Here, on a site that rises from the bay in tiers to a bluff, it founded a village named for Cadwallader C. Washburn, Wisconsin governor from 1872 to 1874. At the natural deep-water harbor, the railroad company and lumber firms quickly installed a dock over which grain and flour from the wheat lands of Minnesota and the Dakotas and lumber from nearby forests would be shipped down the lakes to market. The railroad also built merchandise docks over which coal and supplies from Buffalo and the East could be unloaded and shipped west by rail. Washburn became a connecting point for the transaction of business between the water transportation companies of the East and the Omaha Railroad to the West. Men logged the town site, graded roads, and put in sidewalks. The town boomed overnight. A post office, school, general store, freight warehouse, and elevator, as well as boarding houses, saloons, and sawmills, appeared simultaneously. The village was organized in 1884, and citizens elected its first officers. In 1888, 55 million feet of lumber was cut at Washburn mills, 1,600 cars of Lake Superior brownstone taken from Washburn quarries, and 330 million pounds of flour, merchandise, and coal passed over the docks. The same year, people invested nearly four hundred thousand dollars in physical improvements in Washburn. After an election to decide the issue of location in 1892, the seat of county government was transferred from Bayfield to Washburn.

Washburn State Bank (Washburn County Historical Museum and Cultural Center), Washburn

In 1887, A. C. Probert, formerly head bookkeeper for the White River Lumber Company, established the Washburn State Bank and within two years built a sturdy fireproof brown sandstone bank building on Washburn's main thoroughfare across from the brownstone Union Block, the finest commercial building then in Washburn. Probert announced his intent to construct the brownstone bank building at the very moment the Bayfield County Bank opened for business the doors of its new, modestly Richardsonian Romanesque, fireproof brownstone building. Bayfield brownstone from quarries at Houghton Point three miles from Washburn furnished the material for the Washburn State Bank.

On the morning of 15 September 1888, fire destroyed an entire square block in Washburn, resulting in the loss of nearly thirty buildings. In little more than one year, eighteen new structures rose from the ashes. Among them was the Washburn State Bank.[20] Without doubt, fear of fire precipitated the decision to build the Washburn State Bank in stone.

Conover and Porter of Madison and Ashland drafted plans for the pretentious Richardsonian Romanesque bank in the summer of 1889. Allan D. Conover (1854–1929) and Lew F. Porter (1862–1918) had opened a branch office in Ashland in 1887, the same year they had established their architectural and engineering firm in Madison. Both had studied at the University of Wisconsin; Conover taught civil engineering there and, in fact, instructed and first employed Frank Lloyd Wright. In 1890 English immigrant Horace P. Padley joined the Conover and Porter firm, taking charge of the Ashland office. Conover, Porter and Padley were regarded as the most competent and successful architects in northern Wisconsin, designing such Ashland buildings as the Knight Block (1889), County Jail, County Poor House, Vaughn Block, Been Block, and many houses.[21]

John Halloran of Washburn began construction on the Washburn State Bank in the fall of 1889. The brownstone came from the Washburn Stone Company's quarries, opened in 1885 between Washburn and Bayfield at Houghton Point. Smith and Babcock of Kasota, Minnesota, operated the quarries continuously until at least 1896. The stone is clear purplish reddish brown. Stonecutters finished preparing the stone the first of May 1890, and masons completed laying it in the walls that spring. The roof was tinned in July. The completed project cost nearly twenty-five thousand dollars.

Oriented to a corner site at the northeast intersection of West Bayfield Street and Central Avenue, the rectangular Washburn State Bank measures twenty-five feet in width by eighty feet in length. It is two stories with an attic. Evenly coursed Bayfield brownstone ashlar clads the exterior walls of the main Central Avenue (southwest) and West Bayfield Street (southeast) facades; red brick finishes the walls of the rear (northwest) and side (northeast) walls. Two steeply pitched, tin-covered hipped roofs, the front rising higher than the rear, cover the structure. Hipped dormers project from the roof on all four sides, and tall chimneys thrust through it. The dormers, chimneys, and roofs create a pointed silhouette.

Large rectangular store windows pierce the main facades of the bank at ground level. Above them are round-arched windows, clustered in groups of twos and threes and separated by pairs of rectangular windows. Engaged columns with foliated Byzantine-like capitals run along the southwest and southeast walls of the first story. Resting on the engaged columns, heavy stone lintels mark the first and second stories. Stone steps lead from the sidewalk up to the main entrance, which stands recessed at a forty-five-degree angle at the southeast corner. A lone, free-standing column bisects and supports the main entrance corner. Two side entries give access from Central Avenue to offices at the rear of the bank. Stones inscribed "Bank 1890" are seated over both the West Bayfield Street and the Central Avenue approach to the entrance.

Banking rooms and offices and the offices of the A. A. Bigelow Logging and Lumbering Corporation occupied the first floor. Second and third floors held additional offices. The building was heated with steam and supplied with hot and cold running water.

Two weeks after the Bank of Washburn opened, the *Washburn News* for 17 January 1891 described it in these words:

> The new building erected during the past year by A. C. Probert and occupied by the Bank of Washburn is one of the finest structures of its size in the state and is a model of architectural beauty and fine workmanship. It is constructed of the celebrated Lake Superior Brown Stone taken from our own quarries. . . . The workmanship throughout is of the very finest and the material used is the best that money could buy. It is indeed a building to which every resident of any town only eight years old could point to with just pride, and one which the people of Washburn are proud to have erected in their city.

Unable to pay its depositors in the panic of 1893, the Washburn State Bank closed on 10 June 1893. Officials arrested Probert and charged him with embezzlement. In July 1895 he was sentenced to prison despite his promises to pay off the claims against him. Some said that the bank had been established in the first place for the purpose of defrauding the people.[22]

Today the Bank of Washburn serves as the Washburn Historical Museum and Cultural Center, and a rehabilitation completed in 1993 and

Walker School, 1893–94, T. Dudley Allen, Washburn. Destroyed 1947. From Ashland Daily Press, *Annual Edition (1893).*

a subsequent addition containing the elevator, theater, and exhibit space accommodates this adaptive reuse.

Walker School, Washburn

In the early 1890s Washburn emerged as the center of population, commerce, and government in Bayfield County. The social and economic progress of Washburn and the Chequamegon Bay region ushered into the area an era of maturity. As a manifestation of this cultural progress, in 1893–94, the people of Washburn built on the hilltop above the courthouse square overlooking the village of Washburn and Chequamegon Bay a Richardsonian Romanesque school of rock-faced Washburn brownstone from Houghton Point. The school was named for Lyman T. Walker, president of the high school board. The school was destroyed by fire in 1947.

T. Dudley Allen, a Cleveland architect, designed the Walker School. During his career, Allen designed at least seven courthouses in Ohio, Nebraska, Minnesota, and Iowa. Allen frequently employed in his works the Richardsonian Romanesque in a bold and vulgar way. The Walker School displayed his tendency to mass elements in a seemingly unplanned fashion with the result that the finished work could serve as a courthouse as easily as a school.[23]

In May 1893, as soon as the Washburn School Board secured from the state of Wisconsin a loan of twenty-five thousand dollars for the erection of the new school, it let a contract for its construction to the Minnesota Stone Company of Minneapolis. Costs were estimated as nearly twenty-five thousand dollars. The *Washburn News* for 20 May 1893 reported that the school would be built of Washburn brownstone from the Washburn quarries, and work would begin soon.

Construction on the school progressed sporadically in September, and village and school officials claimed that the builder was not living up to the terms of the contract, refused to pay the firm, and demanded to take the work from the firm and complete the building themselves. When the Minnesota Brownstone Company denied village and school authorities permission to take possession of the building, and an argument resulted in the injury of a foreman, the company sued for damages. Local officials tore down some of the work, and a court injunction was issued. The dispute was resolved by the end of September, and construction resumed under the supervision of superintendents employed by the school board with considerable work being torn down and replaced. In December the Washburn School Board borrowed ten thousand dollars from the state of Wisconsin to complete the school.[24]

The huge sandstone school with parapets, towers, and turrets rose two and one-half stories on a raised foundation to a steeply pitched hip roof.

The first and second floors contained four rooms each. Above the second floor, space for two additional rooms was left unfinished.

The people of Washburn admired their brownstone school because they felt it would benefit the community in several ways. It would renew the enthusiasm of teachers and pupils for their work, demonstrate the far-sightedness of the community, provoke compliments from passers by on trains and steamers, and increase the value of homes in the community.[25]

The completion of the Walker School attracted the attention of people throughout the Chequamegon Bay area. Those in Bayfield noted the recent changes in Washburn and admired the community's progressive leaders. To those in Washburn, the school seemed the finest school, and perhaps the finest public building, in northern Wisconsin.

The Walker School was dedicated the end of May 1894 at the first annual Education Week. For one week the entire community enjoyed entertainment, speeches, concerts, and exhibits that combined commencement exercises with dedication ceremonies. At the ceremony marking the opening of Walker High School, Mary Waegerly, an eighteen-year-old graduating Washburn high school senior, presented an oration on "The Age of Brownstone." She noted in her historical review of education in Washburn that the students were about to enter the "Brownstone Age," giving name to the era of cultural maturity that the people of the region had entered. She found in the native brown sandstone of "the elegant new high school on the hill" a visual metaphor for the era.[26] The brownstone school represented the conclusion of Washburn's pioneer and frontier status and the maturing of life and culture in northern Wisconsin.

Bayfield County Courthouse, Washburn

On 8 November 1892 voters in Bayfield County elected to move the seat of county government from Bayfield to Washburn. Washburn stood nearer to the geographic and population centers of the county than Bayfield, and Washburn businessmen offered to donate a one-block site containing a school building suitable for conversion to county offices and six thousand dollars with which to renovate it. Accepting this offer and building a new jail at Washburn seemed to make better financial sense than expanding the courthouse at Bayfield and constructing a jail there. As soon as the election results proclaimed Washburn the county seat, county records were transferred to temporary quarters in the Washburn Town Hall.[27] Although the current jail's brick walls were substantial, people regarded native Bayfield brownstone as the most durable and beautiful of building materials and knew that building with sandstone supported a local industry. Citizens determined Bayfield brown sandstone to be the only material suitable for public buildings in Bayfield County.

By spring the people of Bayfield County pressed for a new courthouse. Jurors, witnesses, and visitors to the court favored building a new structure over repairing the present county building, and the rapid growth and development of the county soon would require a new courthouse anyway. "Plans for a very elaborate structure, to be in keeping with the new school house and the new jail, will be drawn," predicted the *Ashland Press* for 3 June 1893.

After the election of a new county board of supervisors, the board appointed a committee to investigate matters and deal with the issue of providing suitable offices for the county and suitable safe depositories for the county's records. Dismissing as inadequate the small city-owned wooden structure that the county leased in an old schoolhouse, the committee recommended the board authorize the county clerk to call for plans and specifications for a suitable courthouse. The board acted to receive bids for the construction of "a brownstone courthouse to cost not more that $30,000."[28]

Twelve architects submitted plans for the Bayfield County Courthouse. The Bayfield County Board of Supervisors selected the three best and most practical plans—those by T. D. Allen; Conover, Porter and Padley of Ashland and Madison; and Orff and Joralemon of Minneapolis—and turned them over to a committee for further study. Eventually the board chose the designs prepared by the Minneapolis firm for a structure estimated to cost thirty thousand dollars.

Fremont D. Orff (1856–1914) and Edgar E. Joralemon practiced architecture together in Minneapolis from 1893 to 1897. From 1885 to 1886 Joralemon had worked in Minneapolis, and from 1887 to 1889 was associated in partnership with Charles F. Ferrin. Orff practiced in several partnerships, with his brother George W. Orff (1836–1908) from 1883 to about 1900, with Joralemon until 1897, and with an architect named Guilbert in 1898. From 1899 to 1912 he worked alone, designing houses, schools, libraries, and courthouses. Orff's Minnesota courthouses included the Big Stone Courthouse (1901–2) in Ortonville, the Renville County Courthouse (1902) in Olivia, and the Red Lake County Courthouse (1910) in Red Lake Falls.[29]

Orff visited the site of the proposed courthouse in Washburn. The construction contract was awarded to the lowest bidder, John Halloran of Washburn, for $28,670, and stone for the courthouse came from the quarries of the Washburn Stone Company at nearby Houghton. Halloran contracted with Peter Davidson of Duluth for the brick and stonework.

The *Washburn Times* described the structure as work progressed on 17 April 1895: "The structure will be entirely of Washburn brownstone, with finished surface. There will be several ornamental columns, projecting balconies, etc. A large dome will surmount the structure and will

Bayfield County Courthouse, 1894, Orff and Joralemon, Washburn, photograph 1983. (Kathryn Bishop Eckert)

afford an excellent view of all surrounding country." On 26 June 1895, it continued:

> That dome of the new court house, as it commences to loom high in the air, stands as the apex of a monument of perseverance and triumph in many ways. Aside from these things it marks the nearing of the completion of a piece of architectural and mechanical work that is a credit to all parties concerned in its designing and construction. As a whole the building may be termed a beauty. The classic style of architecture, the carved stones, and all the various intricacies of the mammoth structure present an achievement of beauty and magnificance [*sic*] that an older and richer county than Bayfield could look upon with haughty vanity and supercilious pride.

The Bayfield County Courthouse stands in the center of a full block on East Fifth Street just below the site of the former Walker High School on a plateau overlooking the village of Washburn. The Beaux-Arts Classical public building is built of solid light purplish brown Bayfield sandstone, the rusticated masonry of the foundation and first story contrasting with the smooth-cut walls of the second story. The building is domed. A giant tetra style pedimented Corinthian portico shelters the main entry, which is reached by a broad flight of stairs. Over the entry is a Palladian window with bracketed balcony.

Inside, a central hall runs the full length of the building from southeast to northwest, permitting access to the county offices of the first floor. A divided staircase rises to the second floor.

Later, a very ordinary one-story addition was added to the rear of the courthouse, and in 1998 another more compatible addition was underway. Behind the courthouse is the sheriff's residence and county jail, a brick structure built in 1893 to plans and specifications Number 588 furnished by the Paully Jail Building and Manufacturing Company of Saint Louis, Missouri.

County officials moved into their new quarters in December 1895, and the building was finished early in 1896. Additions were constructed around 1960 and 1997–98.

Free Public Library (Washburn Public Library), Washburn

Four blocks downhill from the courthouse on the corner of Washington Avenue and West Third Street stands the Beaux-Arts Classical Free Public Library. Designed by Ashland architect Henry E. Wildhagen in 1903, the structure's exterior walls are of evenly coursed, rock-faced Washburn brownstone.

In the spring of 1902, a committee of prominent citizens—W. H. Irish, L. N. Clausen, Fred T. Yates, M. A. Sprague, and J. E. Jones, all either judges, newspapermen, or city officials—asked Andrew Carnegie for fifteen thousand dollars with which to construct a library for Washburn. In 1893 the village had built a small, four-room library on Washington Avenue near Bayfield Street and had demonstrated its ability to maintain it. The acquisition of books and the ever-increasing patronage of the institution demanded a larger and more fully equipped building.

On 12 February 1903 James Bertram, secretary for Mr. Carnegie, responded to the request of the Washburn library committee, "If the city agrees by resolution of council to maintain a free public library at a cost of not less that fifteen hundred dollars a year and provide a suitable site for the building Mr. Carnegie will be pleased to furnish fifteen thousand dollars to erect a Free Public Library Building for Washburn."[30]

With this news, the Public Library Board sought a site and secured plans for the proposed library. Named to a site-selection committee were Nels M. Oscar, T. C. McWilliams, W. H. Irish, O. A. Lamoreaux, and C. O. Sowden. The committee identified three possible building sites: the land between the Hotel Washburn and the Bank of Washburn; the Beausoliel corner; and the William Olson property, comprising lots nine and ten of block thirty-nine at the southeast corner of Washington Avenue and Third Street.

Although many preferred the Beausoliel corner because its ample size permitted the placement of the library in a park-like setting, others desired a less expensive location. On 30 June 1903 the board purchased the Olson property for thirteen hundred dollars, more than two thousand dollars less that the asking price of the Beausoliel corner. Nevertheless, the intent of the board to provide a beautiful and permanent library remained firm. In the meantime, the library board had appointed Irish, Oscar, and a Mrs. O'Neill to yet another committee to secure plans for the new building.

The list of functional requirements for the library that would guide the architect first appeared in the *Washburn News* for 21 May 1903. "It is proposed that the main floor will be several steps above the street, and contain several rooms including children's reading room, adult[s'] reading room, reference room, book room and librarian's reference office. This will be all arranged in such a manner that one person can supervise all the rooms, being accomplished by the location of the librarian's office in the center of the building, viewing all the rooms at a glance. There will be a high basement containing a lecture room." The newspaper added that Van Ryin and DeGelleke of Milwaukee, a firm that had drawn the plans for numerous libraries in the state, would probably furnish the plans for a building that met the building committee's program.

Although the building committee logically might have turned to a Wisconsin firm experienced in designing libraries throughout the state, it, indeed, would select a practitioner who was quickly developing a reputation in the Chequamegon Bay area as a designer of public buildings in local stone. The board instructed its president "to engage Architect Henry Wildhagen of Ashland to make plans and sketches of the building at once."[31]

Henry E. Wildhagen (1856–1920) was perfectly capable of giving the people of Washburn an efficient but monumental library. After graduating from the University of Hanover at Hanover, Germany, his birthplace, he immigrated to America at age thirty. In 1893 he opened an office in Ashland with Herman Rettinghaus, a civil engineer. Wildhagen had come to Ashland to design a sulphite mill. His ability to skillfully and carefully draft formal plans and designs in brick, wood, and stone made him the most prominent architect of public buildings in the Chequamegon Bay area at that time. In addition to his own experience in planning schools, courthouses, and libraries, Wildhagen, most certainly, had seen Washburn's public architecture.

Following the earlier models of the Bayfield County Courthouse and the Walker School, Wildhagen and the library board chose to execute the Free Public Library in locally quarried light purplish brown sandstone. When the board opened bids for the construction of the library on 16 January 1904, it discovered with dismay that the only local bidder was not the lowest bidder and that all bids exceeded the fifteen-thousand-dollar Carnegie grant. Yet the board offered the contract to Sheridan and Swain of Washburn because it wanted to favor local people. The selection of a Washburn rather than an Ashland or Chicago contractor expressed the same reliance on, confidence in, and loyalty to local people as had the selection of an Ashland rather than Milwaukee architect. Construction began on 28 April 1904, the cornerstone was laid 9 June 1904, and the library opened and was dedicated in April 1905.

The library measures fifty feet by seventy feet in ground dimensions with a wing at the west (rear) and stands one story on a raised foundation. Large single-light sash windows grouped in threes flank the portico and are placed singly elsewhere. A pair of Ionic columns supports the pedimented central entry portico, and a long flight of stairs climbs to the central entrance. Over the entrance a stone lintel is inscribed with the words "Free Public Library."

The main entry opens to a vestibule and the rooms of the main floor, including the children's reading room, the general or adult room, a reference room, stacks, and the librarian's office. The basement holds lecture rooms and boiler and storage rooms. The people of Washburn viewed the building as "a monument to the conscientious endeavors of the architect, builders and the library. The building is of a beautiful brownstone, taken from the local quarries, and two huge pillars adorn the entrance. . . . It is one of the ornaments to the city which will always be shown with conscientious pride to strangers who visit here."[32]

Ashland

Ashland grew in a large sweeping semicircle on a plain around the southern shore of Lake Superior at the base of Chequamegon Bay. Asaph Whittlesey selected Ashland as a town site in July 1854 when he rowed over from La Pointe on Madeline Island, where John Jacob Astor had established headquarters for the northwest trade of the American Fur Company in 1818. Whittlesey built a cabin on a site at what is now Eighteenth Avenue and Lake Shore Drive. Ashland remained a tiny settlement until its sandstone and timber resources attracted the attention of railroad companies.

Nearly fifty houses sheltered the inhabitants of Ashland in 1857. Within five years, all but the households of the Beaser and the Martin

Roehm Families had left. In 1871–72 new settlers came in anticipation of the arrival of a railroad that would haul out sandstone and lumber. Soon some two hundred buildings were constructed hastily for thirteen hundred workers on the Wisconsin Central Railroad, who made Ashland their headquarters while clearing a track through the forest. The cost of extensive bridge building in the final thirty miles of the road delayed the project, but on 2 June 1877 the Wisconsin Central Railroad finally reached Chequamegon Bay and connected the harbors on this part of Lake Superior with Milwaukee and Chicago. Ashland boomed as brownstone quarries were developed on a large scale, sawmills went up on the shore, and deposits of iron ore described in the 1849 geologic surveys were sought by prospectors in the Penokee-Gogebic Range.

By 1887 four railroad lines terminated at Ashland: the Chicago, Saint Paul, Minneapolis and Omaha; the Northern Pacific; the Milwaukee, Lake Shore and Western; and the Wisconsin Central. The meeting of ship and rail at Ashland made the city an important shipping center. The products from flour mills, a blast furnace, and saw mills at the natural harbor, and iron ore hauled forty miles overland from the Gogebic Iron Range were loaded on ships waiting at the docks. In turn, coal from Ohio and Pennsylvania and merchandise bound for northern Wisconsin, Minnesota, the Dakotas, and Montana were unloaded from ships at the harbor and loaded on trains. By 1900 the population reached ten thousand, but it declined once men had cut over the forests and had failed at attempts to farm the rugged, inhospitable lands.

Union Passenger Depot, Wisconsin Central Railroad (The Depot), Ashland

As early as 1885, the people of Ashland thought the city needed a union railroad station. They hoped the four railroad companies would collaborate and build a centrally located, properly appointed, and sufficiently large depot to accommodate both travelers and businessmen. A native brown sandstone union depot costing about twenty-five thousand dollars would reflect well on the growing city and the railroads. The Wisconsin Central, the Lake Shore, and the Northern Pacific Railroads initially favored building a union station, but the manager of the Omaha Railroad was reluctant to support it.

The Ashland Businessmen's Association's Committee on Transportation discussed the matter of the station with the officials and managers of the four railroad companies. Businessmen and railroad men alike had good reasons for supporting a union station. The businessmen thought Ashland needed a depot large enough to accommodate a city that was expected to double in population in ten years. J. I. Levy stated that a union depot would help trade and make Ashland a more prominent commercial center, postmaster Sullivan indicated it would greatly facilitate mail service, and J. B. Mathews said it would help Ashland's wholesale business by establishing her importance as a trade center with the railroads. C. A. Mitchell of the Omaha Railroad; J. M. Whitman, general manager of the Chicago and Northwestern, of which the Omaha was a part; C. H. Hartly, superintendent of the Lake Shore; E. H. Harrison of the Northern Pacific; and J. S. Dyer of the Wisconsin Central observed that all companies had deferred repairs on their present buildings awaiting a decision on a union station, and that now they should build a better facility.[33]

In November 1887 the railroads reached an agreement under which the Wisconsin Central would build a union station for the common use of all the railroads. To the community the decision seemed to demonstrate respect for the Lake Superior terminus of the railroads and the heavy passenger traffic. The local newspaper happily anticipated that there would be "no more boxes and barns scattered about the city for depots, but an elegant structure such as any city might well be proud of and an honor to the railroad as well."[34] Ultimately the Wisconsin Central Railroad built and owned the station, occupying it with the Northern Pacific, with which it had consolidated interests. The Milwaukee, Lake Shore and Western, and the Omaha never became parties to the depot.[35]

The chief engineer for the Wisconsin Central Railroad brought the plans for the station to Ashland from Chicago, and the newspaper said: "In style it will be Gothic, with a clock tower. The depot proper will be two stories high, and the elevation, including tower will be 144 feet, facing Third street. . . . Every modern convenience will be furnished and the estimated cost is placed at about $45,000."[36] Tentative plans called for a structure of brick and iron.

The plans for the depot were prepared by Charles S. Frost (1856–1931), a prominent Chicago designer of midwestern railroad stations. Frost was educated at Massachusetts Institute of Technology and had worked with Peabody and Stearns in Boston. In Chicago he practiced in partnership with Henry Ives Cobbs from 1882 to 1889 and with Alfred H. Granger from 1898 to 1910. His marriage to Mary Hughitt, the daughter of noted railroad man Marvin Hughitt, brought him commissions for depot designs. Frost designed the La Salle Street Station and the Northwestern Terminal in Chicago, the Union Stations in Omaha and Saint Paul, and the Northwestern Depot in Milwaukee, as well as the smaller stations at Lake Geneva, Superior, Eau Claire, and South Milwaukee, Wisconsin.[37]

Within a year plans for the Ashland station were turned over to local builders Scott and Hubbell to determine the cost of construction, and to the delight of the people of Ashland, the Wisconsin Central asked Scott and Hubbell to estimate the cost of the structure if built of Lake Superior brownstone rather than of brick.

Stone for the station came from either the Ashland Brown Stone Company or the Prentice Brownstone Company. Both companies had recently opened quarries nearby and surely had on hand supplies of roughly cut sandstone easily transportable by water to Ashland. In 1886 John Knight and other Ashland men had formed the Ashland Brown Stone Company and opened a quarry on Presque Isle (Stockton Island). At that time Knight also served as local attorney for the Wisconsin Central Railroad.

That fall Frederick Prentice of New York City had established a brownstone company at Houghton Point, three miles north of Washburn on the Omaha Railroad. Here he owned six hundred acres containing what seemed to be inexhaustible quantities of high quality brownstone. Eighteen months earlier he had begun clearing and stripping the overburden of the site for the quarry. Now a community of one hundred men lived and worked at the quarry site. With five channelers, they cut the solid rock into strips, and with steel wedges driven by sledges wedged them into immense mill blocks that were split into smaller blocks. By this time Prentice had stockpiled a large supply of sandstone. The *Ashland Press* for 20 October 1888 reported: "Huge mountains containing over 60,000 tons of mill stone, stand ready for shipment as rapidly as cars can be secured. Over one hundred and fifty tons of 'ton stone' has also been taken out and is ready for use. A small quantity also has been secured."

As construction was underway in January 1889, both the *Ashland Weekly News* and the *Ashland Press* fully described the station, illustrating their articles with an elevation drawn by the architect. "Iron, wood, brick, and stone will enter lavishly into the construction, and give it not only a maximum of safety and strength, but in the pleasing combination of color, give an effect not readily produced by any other process."[38]

Standing at 400 Third Avenue West, the Richardsonian Romanesque Union Passenger Depot, made of light purplish brown sandstone, measures 158 feet in length and 45 feet in width. On the northwest a central entry tower rises 80 feet. From this tower and the main entrance is an unobstructed view of Chequamegon Bay and the Apostle Islands.

Frost paid careful attention to arrange everything for arriving and departing trains and for the comfort of travelers. On the main floor, a central lobby, two ample waiting rooms, baggage rooms, and dining rooms served passengers. The basement held smoking and rest rooms for men, storage space, and boiler and coal rooms. The second floor contained railroad offices.

When the building was completed and opened to the public, the *Ashland Weekly News* for 28 August 1889 said, "the depot is a beautiful pile of brownstone and brick." The Union Central Depot of the Wisconsin Central Railroad stands as a monument to transportation and its important role in the economic growth and development of Ashland. The rugged brown sandstone station symbolized the hopes of Ashland businessmen for the future of the city as a major shipping center at a time when Superior and Duluth were gaining on Ashland for that position. Today a restaurant occupies the station.

Knight Block, Ashland

In the fall of 1889, in the midst of a building boom during which new commercial blocks and a grain elevator were going up, John H. Knight, a prominent Ashland capitalist and former mayor, began to build a huge commercial block of Bayfield brownstone in the central part of Ashland on the corner of Main Street East and Ellis Avenue. The construction of the Knight Block highlighted the season's building boom.

Knight commissioned the architectural firm of Conover and Porter of Ashland and Madison to draft plans for the three-story block and hotel. As the foundation was excavated, the local newspaper revealed the firm's plans for the building. The large block would have 140 feet of frontage on Main Street East, then named Second Street, and 100 feet on Ellis Avenue, more than any block in the city, and enough space for six stores on the main street. From each street, an entrance and flight of stairs would lead to the upper floors. A central tower would rise above the structure on Main Street East.

The preliminary plan called for the division of the first floor into eight spacious, well-lighted stores, each with its own fireproof vault. It arranged the second floor into offices, all with modern conveniences and eight with fireproof vaults. The third floor held two large social halls with anterooms.

The Knight Block was one of John H. Knight's many business interests. Knight (1836–1903) had moved to Ashland from Bayfield in 1880. At Ashland he founded the Superior Lumber Company, eventual controllers of one of the largest timber tracts in northern Wisconsin. He served as local attorney for the Wisconsin Central Railroad, as organizer of the Ashland and Superior Railroad, and as mayor of Ashland. Born in Kent County, Delaware, and graduated from the Albany, New York, law school in 1858, Knight took a one-year assignment at the Chippewa Indian Agency at Bayfield in 1869 after fighting in the Civil War. Once in Bayfield he worked for several years as register of the land office.

The upper portions of the Knight Block held the Knight Hotel: the third floor had fifty sleeping rooms; the fourth floor contained a dining room that seated 150, a kitchen, wine rooms, a billiard room, and several sleeping rooms; and the tower had four sleeping rooms with a view to Basswood Island and beyond. An experienced staff ran the hotel. To eliminate competition, Knight bought the huge wooden Chequamegon Hotel that fronted on Chequamegon Bay and tore it down.

Knight Block, 1890–92, Conover and Porter, Ashland. Destroyed 1974. **From Ashland Daily Press,** *Annual Edition (1893) 24.*

When the hotel opened on 4 January 1892, the *Superior Daily Leader* called the block "the finest building in Ashland."[39] *Pen and Sunlight Sketches of Duluth, Superior and Ashland,* an illustrated pamphlet published in 1892 that promoted the three Lake Superior communities, described the attributes of the Knight Hotel:

> By all odds the most elegant, most complete and most popular hotel in Ashland, Wisconsin, is the magnificent Hotel Knight, which was first opened to the public Jan. 4th, 1892. This magnificent hotel is located at the corner of Second street and Ellis avenue, in the most central part of the city. . . . The Hotel Knight is practically fire proof, and is strictly first class in every respect. It is furnished with every modern convenience, elevators, baths, electric lights, and has steam heat in every room. There are sample rooms, music hall, wine and billiard rooms in connection, and every appointment is excellent in its details. The building cost over $200,000 and the furnishings were put in at a cost of $20,000. The structure is five stories in height, and is built of brown stone. It measures 140 x 100 feet in dimensions and in design is ornate and massive. The guest

rooms number fifty, and are all furnished equally comfortably; while the cuisine is equal to that of the best hotels in the West. The rates are from $2.00 to $3.00 per day, and every guest receives the greatest care and attention.[40]

In erecting this large sandstone Richardsonian Romanesque commercial block and hotel, Knight proclaimed his position among the people of northern Wisconsin and advertised the product of the Ashland Brown Stone Company, of which he was part owner. The importance of Ashland was evident through the building, and in it the townspeople derived a sense of value and pride. The block was demolished in January 1974.

Ashland Post Office (Ashland City Hall), Ashland

In May 1890 Congress authorized the construction of the United States Post Office at Ashland. Congress appropriated funds for the acquisition of the site and for the construction of the building in August 1890 and March 1891, respectively. Located on the northwest corner of Main Street West and Sixth Avenue West, the Ashland Post Office was designed in the Richardsonian Romanesque style by Willoughby J. Edbrooke, supervising architect of the United States Department of Treasury.[41] It was constructed by Forster and Smith of Sault Sainte Marie, Michigan, under the supervision of Horace P. Padley of Conover, Porter and Padley, architects of Ashland, between August 1892 and May 1894. The exterior walls of the building are of clear Lake Superior brownstone from the Prentice quarry at nearby Houghton Point. The building went up during the years that Ashland vied with Duluth and Superior for a position as a major port on Lake Superior.

In 1889 the Ashland Businessmen's Association promoted with renewed vigor a project discussed five years before by residents of Ashland and northern Wisconsin to secure a public building for Ashland. Competing aggressively with other cities for a federal government building, the association submitted a report to Congressman Myron H. McCord outlining reasons for placing this building in Ashland and describing the commercial importance and advantages of Ashland over other cities. Citizens of Ashland petitioned McCord's energetic and earnest support.[42] Their efforts succeeded; on 12 January 1890 a bill for a public building in Ashland reached the Senate.

Ashland businessmen advanced many arguments for constructing a government building in Ashland. The association argued that the number of federal offices located there and the volume of business transacted through them proved the need. Furnishing suitable quarters under one roof for the Indian agency, collector of customs, register and receiver of the land office, and post office would offer citizens greater convenience and save five thousand dollars in annual rental, eventually enough to pay for the new structure. The businessmen's association felt that a port with over three thousand arrivals and clearances was entitled to police courts and some recognition. Moreover, the association hoped the decision to build a government building in Ashland would lead to the establishment of a federal court in the city to handle conflicts related to problems on the reservation, lakes, at the docks, and trespass on government property. It pointed out the strengths of the prosperous and rapidly growing commercial center of Lake Superior, noting the sharp increase in population from 931 in 1880 to 16,000 in 1889, the capital improvements in 1889 of over $1 million, and the value of commerce in the city for the previous year of $21,257,000.[43]

The bill for a government building called for the acquisition of a site and the construction of a structure costing no more than one hundred thousand dollars in total. President Harrison signed the bill into law on 22 May 1890. It read as follows:

> Be it enacted by the Senate and House of Representatives of the United States of America, in congress assembled, that the secretary of the treasury be, and he is hereby authorized and directed to acquire, by purchase, condemnation or otherwise, a site, and cause to be erected thereon a suitable building, including fireproof vaults, heating and ventilating apparatus, elevator and approaches, for use and accommodation of the United States postoffice and other government offices, in the city of Ashland, state of Wisconsin. The cost of said site and building complete, not to exceed $100,000.[44]

Subsequently, Congress appropriated funds: on 30 August 1890, thirty thousand dollars for the purchase of the site and the commencement of construction; and on 3 March 1891, seventy thousand dollars for the completion of the building.[45]

The secretary of the treasury sought to purchase a centrally and conveniently located lot. Businessmen and property owners on the east and west ends of Ashland competed with each other to secure the location of the government building in their own end of town, and with good reason. In response to the completion and occupancy of the government building, six or more new blocks opened or were planned at or near the intersection of Main Street West and Sixth Avenue West. A special agent for the treasury department inspected over ten sites and examined offers of sale by private individuals. In January the department purchased part of the courthouse park from the county board for six thousand dollars. This parcel included eight lots of the courthouse square—with one hundred feet of frontage on Main Street West, three hundred feet on Sixth Avenue West, and one hundred feet on Lake Shore Drive.[46]

For the next six to eight months, the supervising architect reviewed the space requirements of occupants of the proposed building. Little information on the design and plan of the building was offered to the public with one important exception. An official for the treasury department, Frank Grygla, investigated the qualities of Ashland brownstone as a building material, which led the *Ashland Press* to speculate that the Ashland Post Office would be built of brownstone—Ashland brownstone.

> It is a significant fact that a federal officer, connected with the department which has charge of public buildings, should make a thorough investigation of the merits of Ashland brownstone.
>
> And it is eminently satisfactory when he reaches the conclusion that the brownstone is all, and more than has been claimed. The brownstone industry has already become one of our most important industries, and the market is expanding every year, so that the output will be increased beyond expectation. Not alone is the stone without a peer for a handsome building, but in trimming brick buildings, and for curbing it is unequaled, Mr. Grygla's visit here at this time, also suggests the probability that the $100,000 public building is [*sic*] will be built of this stone.[47]

Grygla's own comments on the stone, in his letter to Edbrooke, were reported by the Ashland paper:

> The stone from the Prentice quarry has been thoroughly tried in many prominent buildings throughout the country. The Milwaukee court house has been built twenty years and can be examined today, with the proof of the durability of the stone. Here in Minneapolis are several magnificent stone buildings put up of the same Lake Superior quarries, and one of them I will name is the Lumber Exchange building, which has been put to a severe test while on fire in the winter of 1891. The building is twelve stories high and the fire burned out all the floors from top to bottom, but the stone structure was left without scaling or checking a particle of the stone work anywhere around the windows, halls and door openings where the flames were pouring out. This should well speak for itself and satisfy the most searching investigation that the stone is capable of resisting any weight put up on it.[48]

On its front page for 19 September 1891, the *Ashland Press* announced that the sketches and plans for the public building had been completed by Edbrooke, reviewed by most of the federal agencies required to do so, and approved by the secretary of the interior. Once the postmaster general examined them, working drawings would be prepared, bids for construction advertised, and work begun. Most importantly, the reporter again noted that the structure probably would be built of brownstone.

Nearly one year later, on 27 August 1892, in an article titled "Of Brown Stone: Ashland's Public Building to be Built of Home Material," the *Ashland Press* confirmed the use of brownstone in the public building. After summarizing the instructions to Horace P. Padley of Ashland, who had been appointed by the treasury department to superintend construction, the article stated: "another important matter which will please every Ashlander, is the fact that the building proper, that is, the facing,—is to be of Prentice brownstone."

On 22 October 1892 the *Ashland Press* fully described the attributes of the local brownstone selected for the Ashland Post Office and optimistically analyzed the role of the rapidly expanding brownstone industry in Ashland's stature as a shipping port. Titled "Our Own Stone: The Government Building to be Built of Prentice Stone: Will be One of the Finest of its Size in the United States," the article lavished praise on stone taken from the Prentice quarry. Significantly, it did not mention the style of architecture, reflecting, perhaps, the rather mechanical, uninspired character of designs for federal buildings generated by the supervising architect, who knew little about the environment in which they would be placed.

> The furnishing of the brownstone for the government building has been awarded to the Prentice Brownstone company and is now being delivered on the ground. It is undoubtedly the finest brownstone that has ever been put into any government building in the United States. It is between a maroon and a cherry color and lights up most beautifully in the sun, works easily, hardens rapidly after it is laid in the wall and does not disintegrate or discolor by age. We think Ashland can pride herself on the finest government building of its size ever built.
>
> The stone is absolutely fireproof, as is shown by the burning of the twelve story building at Minneapolis without even effecting the window, hall or door openings, where the flames were pouring out during the burning of the whole building from top to bottom.
>
> It has since been repaired without removing a single block of stone. Had the building been built of limestone or granite, it would have become simply a pile of broken stone.
>
> Ashland county can boast of more and better brownstone than all the rest of the United States, or, we may say, the world, and it is only a matter of time when this stone is to be shipped to all parts of the United States in immense quantities. It must soon equal the large shipments of lumber now being made from Ashland, and we believe in time will even equal the shipment of iron, and, like the iron shipments, will increase from year to

Ashland Post Office (Ashland City Hall), 1892–94, Willoughby J. Edbrooke, Ashland, photograph 1998. Ashland citizens petitioned the support of their congressman for this federal building and were overjoyed that the supervising architect of the U.S. Treasury Department selected clear Bayfield brownstone from the Prentice quarry at Houghton Point for its exterior walls. (Courtesy Mark Fay, Faystorm Photo)

> year for all time to come, for the supply is practically inexhaustible, and is bound to assist largely in making Ashland the largest shipping port on the great chain of lakes, as the official United States census of 1880 shows that Ashland has already attained third place and is fast crowding Chicago and Buffalo for first place. If the brownstone and iron shipments increase at the rate they have for the past four years, the next United States census will show Ashland the greatest shipping port on the great chain of lakes. This beautiful stone is being quarried along side of and from the same area as the huge monolith for the World's Fair, by the above company.

On 29 June 1892 Edbrooke authorized the postmaster at Ashland to advertise in the local newspapers for bids for the erection and completion, except plumbing and heating apparatus, of the Ashland Post Office. He transmitted copies of the drawings and specifications for the use of bidders in the Ashland vicinity.[49]

With the approval of the acting secretary of the treasury department, Edbrooke accepted on 13 August 1892 the proposal of the lowest bidder, Forster and Smith of Sault Sainte Marie, to furnish the labor and materials for erecting the building for $72,730.

Full instructions from the supervising architect arrived in Padley's office in Ashland on 27 August 1892. They required erecting the building according to that office's plans and specifications, submitting samples of all materials for that office's approval, and using Prentice brown sandstone for facing. The agreement specified that the building be completed within fifteen months.

Work began in August 1892, and the building was occupied in May 1894. As the structure went up, the only discussion of the project in the local newspaper concerned the stone carving. In an article titled, "Carve the Stone," the *Ashland Press* for 31 December 1892 described the process of preparing the material and noted that three-fourths of the dozen stonecutters employed on the job were from Ashland, while the remaining were from Minneapolis.

> The work done on a number of the pieces shows that the building will be a beauty. Large panels six feet by ten, for the front of the building, have been completed. They are artistically dressed, and are very neat. Much of the work is simply plain chiseling of rough blocks for the sides.
>
> There is a large amount of stone piled up in the court house yard to be fashioned over, and the workmen say that by working full force every day, they cannot finish it before June 1st. Work will be begun in the early spring laying the blocks. According to contract the whole thing must be done before Sept. 1st. The foundation of the building is entirely completed now, and the brownstone wall is finished about six feet in hight [*sic*] above this.

The Ashland Post Office measures fifty feet in width by eighty feet in depth and stands two and one-half stories on a raised foundation. A slate-clad roof of intersecting gables pierced by dormers covers the building. Rock-faced random ashlar is laid in the exterior walls. Large round-arched windows and entrances from Main Street West and Sixth Avenue West on the first story give way to smaller paired rectangular windows on the second story and still smaller arched windows in the attic and tower. From the main mass at the southeast corner, a square tower with a round-arched open belfry with balustrade rises to its pyramidal roof. Carved foliation decorates the capitals of the short clustered columns that flank the main entrance at Sixth Avenue West. A carved eagle overlaid with the initials "US" and carved foliation fills the peak of the main east gable.

The post office occupied the entire first floor, with public lobbies approached through entrances on Main Street West and Sixth Avenue West. Officials of the United States Land Office, the Indian agency, and the collector of customs were housed in the seven offices of the second floor. The third story contained five rooms for other offices. At the southeast corner stairs rise through the elaborately paneled tower from the main floor to the upper floors and the tower room itself.

A one-story, flat-roofed addition for a federal court at the rear (north) of the structure is of similar stone. In 1977 the government building was adapted for reuse as the Ashland City Hall. The Ashland Post Office shows restraint in its design. This may be due to the cumbersome process by which it was created and the limited public role in that process.[50]

In its annual edition for 1893, the *Ashland Press* commented on the city's public buildings and noted: "The city of Ashland has one of the most beautiful public buildings in the State of Wisconsin. Congress recognized the importance of the 'Garland City of the Inland Sea' with an appropriation of one hundred thousand dollars for the erection of a public building in Ashland. . . . According to the specifications as laid down by Uncle Sam, all materials are the best that money can buy. . . . The new building is built of native brownstone, procured from quarries adjacent to Ashland."[51]

Produced by a lengthy and complex federal process, the Ashland Post Office is a simple, but somewhat mechanical, statement in local brown stone of the federal presence in Ashland.

Superior

Superior grew on the southern shore of Lake Superior's western tip as a shipping port for the iron ranges of Minnesota and the agricultural lands of the Northwest. The city extends across a flat plain at the confluence of the lake on the northeast and the Saint Louis River on the northwest. Duluth lies across the lake and river and rises to a range of hills. Two long thin points of land formed by the waves of Lake Superior acting against the silt carried here by the river extend toward each other forming a natural harbor. Here grain elevators, coal cranes, and ore docks were the hub of the activities associated with the loading and unloading of lake freighters and trains.

Although French explorers, fur traders, and missionaries arrived in the seventeenth and eighteenth centuries, the first permanent settlers came to Superior after the signing of the Treaty of La Pointe in 1854. The treaty erased Indian title to the region. Then the completion of federal land surveys, news of the building of the locks at Sault Sainte Marie, and the prospect of a railroad from the head of Lake Superior to Puget Sound in the state of Washington led to a vision that Superior would become a transportation, commercial, and industrial gateway. Prominent American businessmen, financiers, and politicians collaborated in promoting and developing a town site at Superior and in seeking federal appropriations for harbor improvements, public surveys, and a railroad to the West. Then R. R. Nelson, Judge D. A. J. Baker, and Colonel D. A. Robertson of Saint Paul, later joining forces with Henry M. Rice, also of Saint Paul, claimed land at the western end of Superior. Soon southern bankers and politicians—Washington, D.C. banker William W. Corcoran, Mississippi Senator Robert J. Walker, Kentucky Congressman John C. Breckenridge, and others—speculated in land and developed the eastern portion of Superior.

Superior boomed when rumors of proposals for railroads to Superior stirred. Although a railroad reached Duluth in 1870, the first railroad, the Northern Pacific, did not arrive at Superior until 1881. The panic of 1857, the Civil War, and an economic decline in the 1870s slowed the project, but Superior experienced a rush of activity by the 1880s, when the construction of the Northern Pacific was completed from Puget Sound to Superior. In 1884 the first iron ore was shipped from the Gogebic Range and, a year later, from the Vermillion Range in Minnesota, and business activity at Superior flourished.

In the midst of this activity, New York and Saint Paul financiers, headed by John Henry Hammond, platted West Superior in 1885 and offered lots to prospective settlers to induce them to come. Scandinavians, Finns, Canadians, Germans, and Poles arrived to work as lumbermen, miners, and stevedores. Within five years, grain elevators, docks, warehouses, commercial blocks, schools, and churches went up. Between 1890 and 1900, Superior was the second largest city in Wisconsin. It shipped and stored millions of tons of ore and millions of bushels of grain yearly. But the development of the Duluth harbor offered more direct access from the Mesabi Iron Range to the water, and Duluth, not Superior, became the major shipping center of the Lake Superior region.

Minnesota Block (Board of Trade Building), Superior

In the early 1890s the Land and River Improvement Company of Superior invested $360,000 in the construction of three commercial blocks in Superior. All were named for the places of origins of the community's early investors and promoters. The Minnesota, Washington, and Berkshire Blocks stood in the commercial center of Superior.[52] The finest of these was the Beaux-Arts Classical Minnesota Block, designed by Charles Coolidge Haight of New York City.

Minnesota Block (Board of Trade), 1891–92, Charles Coolidge Haight, Superior, photograph 1998. (Courtesy Bernard E. Benson)

John Henry Hammond (1833–90) organized the Land and River Improvement Company to quickly develop "an obscure frontier village into a modern city" between 1886 and 1892. An experienced city and railroad builder, Hammond assessed the prospects of creating a city at Duluth and Superior and concluded that West Superior would be the most desirable location. As soon as the city was platted in 1886, Hammond organized and began managing the Land and River Improvement Company. Despite the reservations of some about building a city in a swamp, Hammond succeeded in attracting speculators and investors to the company because of his attention to planning and to the execution of details. The company laid water mains, dug sewers, paved streets, erected docks and modern buildings, and encouraged industries to locate there. Besides advancing the development of the town site, Hammond promoted the Northern Pacific Railroad; the Eastern Railroad of Minnesota; the Duluth, South Shore, and Atlantic Railroad; and the Lake Superior Terminal and Transfer Company.[53]

Hammond and his collaborators selected Charles Coolidge Haight as the architect for the Minnesota Block. Haight (1832–1917) was steeped in the Gothic work of Richard Upjohn. Haight prepared himself to practice architecture by studying in an architect's office and traveling abroad. In his New York City architecture practice, he designed many Gothic churches and college buildings.[54]

F. A. Fisher and Company began building the Minnesota Block in the summer of 1891 and completed it by September 1892 at a cost of

$225,000. James C. Dawkins, a dry goods store, occupied the basement and ground floors. The Land and River Company and the West Superior Iron and Steel Company took some of the upper floor office space.

The Minnesota Block stands at 1501-1511 Tower Avenue on the southwest corner of Belknap Street, once a major intersection in the city. It is just north of a square that contained the Massachusetts Block and the LaSalle brick and cream terra cotta Washington Block. Then under construction at Tower Avenue and Winter Street was the Kasota limestone Berkshire Block. The *Superior Daily Leader* remarked that "the Minnesota, Berkshire and Washington Blocks, now well on the way toward completion form a trio of magnificent business blocks that would do credit to a city older and more pretentious than Superior."[55]

The four-story Minnesota Block extends for 150 feet on Tower Avenue and for 130 feet on Belknap Street. Its cage construction was made fireproof with tile arches laid over iron girders and posts, floors laid on a cement bed covering the tile arches, and outer walls of brick and stone.[56] Light yellowish pink Illinois pressed brick rises in the exterior walls to the ample copper cornice oxidized green. Portage Entry red sandstone trims the exterior. Large plate glass windows pierce the walls. The interior was finished in quartered oak, and marble lined the stairways. The Minnesota Block exemplifies the less frequent use of solid Lake Superior red sandstone by builders in Superior than by builders in Duluth and elsewhere in the region of the Jacobsville formation and the Bayfield group in favor of brick trimmed with sandstone. The fact that Superior was built on speculation may be the reason. Yet the Minnesota Block was said to present an appearance of "solidity and strength combined with symmetry and elegance."[57]

T. W. Marble Double House, Superior

An interesting red sandstone residence in Superior is the elegant house built in 1891 by T. W. Marble at 1311 John Avenue. This substantial building and similar residential blocks then going up all over the city were termed "monuments to the prosperity and solid growth of the city."[58] The three-story double house with a mansard roof measures forty-five feet by fifty-feet in plan. The exterior walls are light yellowish tan brick with a solid brownstone front. Two distinctively different picturesque towers, one with a pyramidal roof, the other with a hexagonal roof, rise above the full height of the building at the center of each house unit, marking each as an individual dwelling. A one-story wooden porch shelters the front entries. The house cost fourteen thousand dollars. The design resembles those created by Barber and Barber, architects of Superior, and could have been built just as easily in Marquette, Houghton, Hancock, Ashland, or Duluth as Superior.

Trade and Commerce Building (Superior City Hall), 1890, Clarence H. Johnston, Superior, photograph 1998. (Courtesy Bernard E. Benson)

Trade and Commerce Building (Superior City Hall), Superior

In 1890 Henry D. Minot, president of the Eastern Railway Company of Minnesota, an affiliate of the Great Northern Railroad, which had laid a track from Saint Paul to Superior, built the Trade and Commerce Building at the northwest corner of Broadway and Hammond Avenue. Minot was a leading promoter of the development of Superior.

Minot commissioned Clarence H. Johnston of Saint Paul to design the Trade and Commerce Building for use as a grain exchange. Johnston (1859–1936) studied architecture at Massachusetts Institute of Technology and in 1886 opened a practice in Saint Paul. There he designed many public buildings, churches, and houses. In 1901 the Wisconsin Board of Control retained Johnston, and he drew plans for many state buildings.[59]

Minot awarded the contract for the masonry work to Ring and Tobin of Superior, the iron work to Herzog Brothers of Minneapolis, and the fireproofing to the Pioneer Company, also of Minneapolis. Construction began in May 1890, and the building was completed one year later.

The three fronts of the Richardsonian Romanesque building were handsomely finished with pinkish yellow Southern Potsdam sandstone. The huge rough-cut ashlar blocks of the exterior walls diminish in size as the building rises its five full stories.

The Trade and Commerce Association leased the entire building. With an illustration of the substantial stone building, the association advertised for memberships and the leasing of office space in the structure as though the prestige and security of the organization occupying space within was reflected in that of the building. Many predicted the Trade and Commerce Building would serve as a model for other good buildings in Superior. Stately and beautiful, substantial and graceful, this structure would "educate the eye and taste of the people."[60] The Trade and Commerce Building seemed to stand "alone in its magnificence" as a symbol of Minot's faith in the future of Superior. Seemingly oblivious to the red Lake Superior sandstone to the east, however, Minot and Johnston chose a lighter sandstone from the nearby formation.[61]

Henry D. Minot's untimely death in 1891 in a train accident terminated plans to use the building as a grain exchange. Subsequently, the building was subleased and leased to the court, library, and offices, but never was fully occupied. In 1904 the city acquired it for only half the cost of construction and used it for offices until 1970.[62]

Duluth

At the head of Lake Superior, Duluth rises on granite bluffs six to eight hundred feet above the lake. The city extends for miles along the shore of Lake Superior, the Duluth-Superior Harbor, and the lower reaches of the Saint Louis River.

An Ojibway village stood at Fond du Lac in 1630. Radisson and Grosellier explored the area in 1654–60. Daniel Greysolon, Sieur du Lhut, for whom the city is named, first visited the area in 1679. From the seventeenth-century explorations until 1847, people here were concerned primarily with fur trading. The arrival of George P. Stuntz in 1852, under the orders of Surveyor General George B. Sargent, and his enthusiasm for the region brought permanent settlement.

After the Treaty of La Pointe, which was signed at Fond du Lac, the Grand Portage and Fond du Lac Indians relinquished their rights to mineral tracts in the region. Rumors of copper and iron deposits and the anticipation of the opening of the locks at Sault Sainte Marie attracted prospectors and land speculators to the area in 1854–55. In 1856 the village of Duluth was established and designated the seat of county government. Jay Cooke's

General view of Duluth from the rocky bluffs with harbor in distance, photograph date unknown. (Courtesy Library of Congress)

promise to make Duluth the northern terminus of the Lake Superior and Mississippi Railroad, later the Saint Paul and Duluth Railroad, and still later, a component of the Northern Pacific Railroad system, brought additional people here. From its harbor were shipped grain, iron ore, coal, and lumber. In 1886 the city had railroads, elevators, docks, and sawmills, and the population reached twenty-six thousand. This doubled in 1900 and grew to one hundred thousand in 1940. Duluth is the third largest city in Minnesota.

The harbor is protected by two narrow sandbars—Minnesota Point and Wisconsin Point—that extend for miles. The points form a basin, and between them are channels of the Saint Louis and Menadji Rivers as they flow into Lake Superior. The outlet and a canal dug through Minnesota Point, improved by Corps of Engineers bulkheads, piers, and breakwaters, are the Superior Ship Canal. In 1893 this became the Duluth-Superior Harbor.

Duluth Central High School, Duluth

The sandstone architecture of the Lake Superior region culminates at the head of the lake in the great Duluth Central High School of 1891–92.

The Duluth Central High School is one of the most prominent sandstone structures in the region. Designed by Emmet S. Palmer and Lucien P. Hall of Duluth, the massive and elaborate high school itself resembles H. H. Richardson's Allegheny County Courthouse and Jail (1884–87) in Pittsburgh.

Construction on the Duluth school began in June 1891 and was completed 15 December 1892 at a cost of five hundred thousand dollars. The Duluth Central High School is built of solid purplish brown Bayfield sandstone.[63] The high school occupies an entire city block on a hillside bounded by East Second Street, First Avenue East, East Third Street, and Lake Avenue. Rectangular wings with steeply pitched roofs and tall protruding dormers flank the central mass. The central clock tower soars to extreme height to match the rugged hillside the building stands against in a proclamation of the value of education to all who enter the harbor. Three Syrian arches flanked by polished granite columns mark the entries. The exterior walls are encrusted with carvings of foliated ornamentation, cherubs, and grotesque animals. The school held fifteen hundred students. Noteworthy were its auditorium, manual training shops, and physical laboratory. The Duluth Public Schools continue to use the central school for educational and administrative purposes. An alternative program and a charter school hold classes in the building, and the Duluth School District has administrative offices on the second floor.

Duluth Civic Center, Duluth

The Duluth Civic Center is a public square defined on three sides by three monumental Beaux-Arts Classical public buildings and open toward the lake according to the tenants of the City Beautiful Movement. Built of sparkling white stone, the complex of three buildings is made up of the Saint Louis County Courthouse by Daniel H. Burnham and Company (1908–9), the Duluth City Hall by Thomas J. Shefchik (1928), and the Federal Building by federal architects (1928–30). This white stone civic complex and similar but smaller civic complexes in Superior and Ashland mark the end of the picturesque and imaginative building with reddish and purplish brown Lake Superior sandstone in northern Wisconsin and Duluth.

Summary

The Chequamegon Bay and south shore areas of Lake Superior in northern Wisconsin and Duluth grew when its ports became gateways of commerce and industry after the locks at Sault Sainte Marie opened and after the Northern Pacific Railroad connected Lake Superior with Puget Sound.

Although somewhat less rugged than Marquette and the Copper Country in its physical appearance, the area possesses an equally harsh climate.

After engineers charted the region's harbors and scientific surveys identified the region's timber and sandstone resources and nearby iron ore deposits, eastern and urban investors William F. Dalrymple, John Henry Hammond, and Henry M. Rice formed companies, such as the Bayfield Land Company and the Land and River Improvement Company, to develop commerce at the ports of Duluth, Superior, Bayfield, Ashland, and Washburn. Soon grain elevators, flour mills, blast furnaces, sawmills, and iron ore, coal, and merchandise docks sprang up. The surge in settlement brought a need for not only industrial buildings but other kinds of buildings. The architects, builders, and clients frequently used the locally extracted sandstone from the quarries in the Bayfield group. The stone gave stability to the architecture, the institutions they housed, and the communities in which they stood. These sandstone buildings belonged to their environment and advertised the product of the local quarry industry. Architects and clients alike recognized the strong local support for the use of Bayfield sandstone.

Since trained architects did not live in the Chequamegon Bay area until the 1890s, and those in Superior apparently did not satisfy the requirements of the more worldly clients, eastern and urban clients commissioned architects from the cities of their origins to create sophisticated designs for stone buildings in the Chequamegon Bay and south shore areas. Seeking to promote the development of Superior, John Henry Hammond, president of the Land and River Improvement Company, selected Charles Coolidge Haight of New York to draft plans for a large, fireproof, Beaux-Arts Classical commercial block of brick trimmed with Portage Entry sandstone and built it at Superior in 1891. Hoping to develop Superior into a center of the grain trade, Henry D. Minot of Saint Paul chose Clarence H. Johnston—trained at Massachusetts Institute of Technology and a practicing architect in Saint Paul—to plan a tall and substantial office building of pinkish yellow sandstone for use as a Trade and Commerce Building. These large substantial stone and stone-trimmed structures demonstrated the investor's belief in the future growth of the city. The United States Department of the Treasury's supervising architect Willoughby J. Edbrooke created a standard Richardsonian Romanesque design for the Ashland Post Office in 1893, and, at the insistence of local citizens, the post office achieves a local identity through the use of purplish brown Ashland sandstone.

Local governments sought the services of moderately trained and talented midwestern architects to draft designs for their public buildings. Thus, the Bayfield County Board of Supervisors chose John Nader, an engineer and surveyor of Madison, Wisconsin, to design its brownstone courthouse at Bayfield in 1886. And, only eight years later, after the seat of county government was moved from Bayfield to Washburn, the board selected Orff and Joralemon of Minneapolis to design the Beaux-Arts Classical courthouse at Washburn. In both cases, the extensive use of locally quarried sandstone served to establish and maintain the status quo of county government. In the courthouse at Bayfield, the board, influenced by local businessmen, employed sandstone to advertise the area's developing quarry industry. In the courthouse at Washburn, the board used Washburn sandstone to meet a primary requirement of the community's definition of a public building and to confirm the location of the county seat.

When the Washburn School Board planned the Walker School in 1892–93, it called on schoolhouse and courthouse designer T. Dudley Allen to create a towered and turreted Richardsonian Romanesque Washburn sandstone structure. The size, location, and material of this school more than any other structure established the standard in Washburn for a public building. Ten years later, the Washburn Library Board selected Henry E. Wildhagen, a formally trained German immigrant, to design a Beaux-Arts Classical Carnegie library. The people of Washburn insisted that the library meet its expectations for a public building and built it of native sandstone.

In 1889 John H. Knight and A. C. Probert commissioned Conover and Porter of Madison and Ashland to design the Knight Block and the Bank of Washburn, respectively. This firm created two flamboyant stone structures, demonstrating the affinity of local clients and local architects for the local material.

As much at harmony with its surroundings as any structure was the classical National Bank Building built at Bayfield in 1905. By then, the quarry industry had declined, and anywhere else a building of this style and at this time would have been of white stone. At Bayfield, Archie Donald, a skilled Ashland stone mason, crafted the bank building of Bayfield sandstone from the nearby Pike quarry.

The businessmen who took pride in local growth and promoted local development realized the importance of building with local sandstone. In nearly every case, businessmen and chambers of commerce greatly influenced decisions of outsiders and local people alike to build in sandstone. Thus, sandstone architecture in northern Wisconsin and Duluth conveyed solidity and stability and promoted development and cultural achievement.

APPENDIX 1

SIGNIFICANT QUARRY COMPANIES IN THE JACOBSVILLE FORMATION AND THE BAYFIELD GROUP

Company Individual	Capital Stock	Stockholders	Main Office	Incorporation Date	Quarry Locations	Comments
Amnicon River Quarry					S32 T48N R12W	
Arcadian Brownstone Co.		A. J. Cheney W. S. Maxwell	Kenosha, WI	1886	S32 T48N R12W on spur of Northern Pacific Railroad and Amnicon River, Douglas County, WI	W. S. Maxwell, manager 1888—first shipment 1889—crew of thirty working; village of boarding houses built for workers 1897—still operating
Ashland Brown Stone Co.	$50,000	John H. Knight William Knight D. S. Kennedy	Ashland, WI	1886	Lots 1, 2, 3, 4 S4 T51N R2W south shore of Stockton Island, Ashland County, WI	Briefly in 1871, Tyler and Healy extracted rubble from site for Willard, Mercer and Co. of Duluth William Knight, general manager and superintendent 1886—quarry opened and stone extracted 1887—extracted 25,000 cu. ft. 1890—sold to J. G. Bodenschatz
Ashland Brown Stone Co.	$100,000	G. A. Bodenschatz J. G. Bodenschatz E. H. Brown	Chicago, IL	1890	S4 T51N R2W south shore of Stockton Island, Ashland County, WI	Frank Bell, foreman 1891–92—shipped 200,000 cu. ft. 1893—shipped 175,000 cu. ft. 1894—shipped 230,000 cu. ft. 1895—shipped 285,000 cu. ft. 1896—shipped 240,000 cu. ft. 1897—suspended operations
Bass Island Brown Stone Co.		Alanson Sweet Daniel Wells	Milwaukee, WI	1868	Lots 1 and 2 SE1/4 SW1/4 S4 T50N R3W south end of Basswood Island, Ashland County, WI	Furnished stone for the Milwaukee County Courthouse

Company Individual	Capital Stock	Stockholders	Main Office	Incorporation Date	Quarry Locations	Comments
Bass Island Brown Stone Co. (Strong, French and Co.)		Edwin C. French George P. Lee Robert H. Strong Daniel Wells	Milwaukee, WI Chicago, IL	1870	SE1/4 SW1/4 S4 T50N R3W south end of Basswood Island, Ashland County, WI	Except for Wells, investors were Chicago stone merchants 1871—quarry operation initiated; Mr. Moneghan, superintendent; 60–70 workers employed 1879—leased by new company
Bass Island Brown Stone Co.	$45,000	Walter C. Brooks Cambreton Leach James S. Ritchie	Superior, WI	1890	Lot 1?, 3?, SE1/4 SW1/4 S4 T50N R3W Basswood Island, Ashland County, WI	Leased site from Breckenridge heirs Blasted out stone but abandoned work soon after beginning
Bayfield Brownstone Development Co.	$50,000	Francis W. Denison Robinson D. Pike George H. Quayle	Bayfield, WI	1892	Stockton Island, Ashland County, WI Tassell Point, Bayfield County, WI	William Knight, manager
Borgeault Stone Co.				1897	S27 T49N R9W on Iron River, Bayfield County, WI	Leased site from Iron River Red Sandstone Co.
Breckenridge Quarry		John Breckenridge	Kentucky	1892	Lot 3, S4 T50N R3W east shore of Basswood Island, Ashland County, WI	Blasted out stone but abandoned work after two or three shipments
Burt Free Stone Co.	$500,000	A. Judson Burt Hiram A. Burt John Burt William Burt William A. Burt	Marquette, MI	1872	S26 T48N R25W South Marquette, Marquette County, MI	Idle by 1879 1883—leased site to John H. Jacobs
Building Stone and Mineral Exploring Co.		F. Bending C. H. Call V. B. Cochran W. S. Hill J. H. Jacobs John Mitchell B. S. Packard R. W. Powell Gad Smith	Marquette, MI	1891	Adjoins Furst, Jacobs and Co. property (Lot 1, S19 T53N R32W) Houghton County, MI	1891—contracted with Powell and Mitchell to strip its quarry site at Portage Entry

Butler Brownstone Co.	$300,000	George Barnes Thomas Butler Charles Johnston F. M. Moore C. H. Schaffer	Marquette, MI	1891	S14, 15, 22, 23 T48N R21W Rock River, Alger County, MI	Quarry well equipped by 1892 Probably closed by 1895
Cook and Hyde	$300,000	Thomas D. Cook Edwin Hyde	Milwaukee, WI Minneapolis, MN	1883	Lot 1, SE1/4 SW1/4 S4, S27 T50N R3W Site adjoining north side of Bayfield Brownstone Co., Basswood Island, Ashland County, WI	1883–88—leased Basswood Island site from R. W. Lee of Chicago 1886—operated site north of Bayfield Brownstone Co.; 100 workers employed at both sites 1889—sold machinery and equipment from both sites to Prentice
Cranberry River Quarry		Omeis F. H. Quinby		1891–92	S5, 8 T50N R7W Cranberry River, Bayfield County, WI	Operated for six months, then Lake Superior Brownstone Co. acquired property
Peter Crebassa		Peter Crebassa	L'Anse, MI	by 1881	S25 T51N R33W Baraga County, MI	
Detroit and Marquette Brownstone Co. (Davids' Quarry) (Detroit Brownstone Co.)	$500,000	A. A. Albrecht Joseph H. Berry Thomas Berry John H. Bissell H. W. Candler I. Candler W. R. Candler M. W. David Myron E. David O. I. David I. G. Hollands H. Hubert Humphrey Alfred Kidder A. L. Lane Leonard Laurence Charles H. Little Willard S. Pope I. C. Smith	Detroit, MI	1889	S35 T48N R25W Mount Mesnard, south slope, Marquette County, MI	1874—quarry first opened by T. T. Hurley 1889—Mark J. David, superintendent; boardinghouse, office, blacksmith shop, stable, and food house built at quarry 1892—renamed Detroit Brownstone Co. 1893—ceased operations, sold personal property and leased quarry to others 1913—sold all remaining property

Company Individual	Capital Stock	Stockholders	Main Office	Incorporation Date	Quarry Locations	Comments
Duluth Brownstone Co. (J. H. Crowley and Co.)	$30,000	John H. Crowley Frank B. Lazier Alexander McDougall	Duluth, MN	1887	Lots 1, 2, S6 T48N R15 W Fond du Lac, Douglas County, WI	Number two or three grade stone Extensive local market in Superior and Duluth but also sold in Minneapolis and St. Paul
Excelsior Brownstone Co. (Prentice Brownstone Co.)	$1,800,000	Edwin Ellis S. S. Fifield F. E. Goddard L. C. Tobias Frederick Prentice	Ashland, WI New York, NY	1893 1892	SE1/4 S13 T51N R3W southeast side of Hermit Island, Ashland County, WI	1880—E. E. Davis of Ashland prospected for quarry location on property Frederick Prentice had owned since 1857 1891—in operation 1892—produced 150,000 cu. ft. 1893–95 produced 220,000 cu. ft. 1896—produced 100,000 cu. ft. 1897—experienced financial difficulties and ceased operations
Excelsior Red Stone Co.	$50,000	Otis W. Johnson Frank E. Robson James B. Seager James H. Seager Samuel L. Smith	Hancock, MI	1891	S8 T53N R32W two miles from Red Rock, Houghton County, MI	James Seager, manager of quarry; Fred L. Smith, general manager in Detroit Stone transported by rail from quarry to Red Rock and shipped over dock of Kerber-Jacobs Co. 1894—shipped 18,000 cu. ft. 1895—shipped 45,000 cu. ft. 1896—ceased operations
Flag River Brown Stone Co.	$30,000	Miller	Duluth, MN	1889	S19 T50N R8W Flag River near Port Wing, Bayfield County, MI	Leased site from Isaac Wing and operated it until 1894 Portage Entry Quarries Co. worked here until early 1900s 1891—extracted 100,000 cu. ft.

Furst, Jacobs and Co.	$180,000	George C. Furst Henry Furst Ernst Heldmaier John H. Jacobs Peter W. Neu Peter A. Pickel	Chicago, IL	1887	S26 T48N R25W South Marquette, Marquette County, MI Lot 1, S19 T53N R32W Portage Entry, Houghton County, MI	1885—produced 90,000 cu. ft 1887–91—produced 300,000 cu. ft. 1890—production peaked at 450,000 cu. ft. 1891—John H. Jacobs sold out interest, and company restructured as Furst, Neu and Co.
Furst, Neu and Co.		Henry Furst Peter W. Neu	Chicago, IL	1891	S26 T48N R25W South Marquette, Marquette County, MI Lot 1 S19 T53N R32W Portage Entry, Houghton County, MI	Furst and Neu were cut-stone contractors in Chicago 1892—extracted 228,643 cu. ft. block stone; employed seventy-five 1893—consolidated with Portage Entry Red Stone Co. as Portage Entry Quarries Co. By 1895 J. W. Wyckoff was superintendent
Furst, Neu Co.	$35,000	Howard M. Carter Henry Furst William Garnett, Jr. Ernst Heldmaier Anna M. Neu Peter A. Pickel Ira C. Wood	Chicago, IL	1899		Corporation formed to quarry and deal in building stone
Grand Island Quarry		Schoolcraft Furnace Co.		1866	SE1/4 S15 T47N R19W southeast end of Grand Island Alger County, MI	Floated stone across bay for construction of Schoolcraft Furnace
Hancock Sandstone Land Co.		James Seager	Hancock, MI	by 1888	Near Portage Entry on north side of lake, Houghton County, MI	
Hartley Bros.				1885	S27 T49N R4W Bayfield County, WI	1885—opened unsatisfactorily 1887—opened at nearby site 1889–92—produced over 300,000 cu. ft. 1897—idle

Company Individual	Capital Stock	Stockholders	Main Office	Incorporation Date	Quarry Locations	Comments
Hennessey Quarry				1860	S25 T41N R33W Baraga County, MI	
Ingalls Quarry		Edmund Ingalls	Duluth, MN	1869	Lots 1, 2 S6 T48N R15W on bluffs next to Saint Louis River at Fond du Lac, Douglas County, WI	1869—quarry opened 1873—ceased operations
Iron River Brownstone Co.	$150,000	J. H. Horton R. I. Tipton Albert C. Titus	Superior, WI	1893	S4? T49N R9W one mile above mouth of Iron River, Bayfield County, WI	Opened in 1892 and soon transferred holdings to Iron River Red Sandstone Co.
Iron River Red Sandstone Co.	$100,000	T. V. Badgley S. B. Roberts A. C. Titus	Superior, WI	1894	S4? T49N R9W one mile above mouth of Iron River, Bayfield County, WI	Stone shipped by scow to West Duluth, then by rail or water to other markets Operated until 1897, then leased for one year to Bourgeault Stone Co.
Kerber-Jacobs Redstone Co.	$1,000,000	Carl Findeisen S. W. Goodale J. H. Jacobs John Mitchell B. S. Packard D. W. Powell E. H. Tower	Chicago, IL Marquette, MI	1892	S8 T53N R32W Portage Entry, one to two miles from Jacobsville at Red Rock, Houghton County, MI	W. J. Fales, superintendent; John H. Jacobs, president and general manager 1893–95 produced over 200,000 cu. ft. 1896—market dull, but shipped 125,000 cu. ft. 1897—Jacobs sold out
Keweenaw Redstone Co.		D. L. Case Theo. W. Luling	Marquette, MI	1891	S22 (or 27) T52N R33W Newton homestead, Newtonville, Baraga County, MI	Leased 160 acres of Newton homestead
Lake Linden Brownstone Co.			Chicago, IL	by 1885–89	Thomas A. Trevathan Farm on DSS&A railroad six miles east of Houghton, Houghton County, MI	Until 1889 worked sporadically on small scale then leased site to Jacob and Israel Schureman
Lake Superior Brown Stone Co.	$500,000	Varnum B. Cochran William S. Dalliba Alfred Kidder Egbert J. Mapes Jay C. Morse Henry Stafford	Marquette, MI	1873	Lot 1, S19 T53N R32W west side of Keweenaw Bay and north side of Portage Entry, just east of lighthouse reservation, Houghton County, MI	1883—leased site to Wolf and Jacobs Co., who developed it

Lake Superior Redstone Co.	$2,000,000	George Barnes C. H. Call D. W. Powell Ernest Rankin Gad Smith	Marquette, MI	1891	Lot 3, S19 T53N R32 W Jacobsville, Portage Entry, Houghton County, MI	Company formed in 1889, after Building Stone and Mineral Exploring Company explored site; leased 55 acres 1896—sales weak, and operations ceased
L'Anse Brown Stone Co.	$500,000	Edward Breitung Timothy T. Hurley Francis M. Moore Paul Pine	Marquette, MI	1875	S25 T51N R33W on L'Anse Bay 1.75 miles northeast of L'Anse, Baraga County, MI	E. M. Wood and George Craig, managers 1876–78—produced nearly 30,000 cu. ft. annually 1879—went bankrupt and sold to St. Clair Brothers of Ishpeming for $5,000
L'Anse Brownstone Co.		W. A. Amberg William O'Brien and other Chicago parties	Chicago, IL	1893	S25 T51N R33W on L'Anse Bay 1.75 miles and other Chicago parties northeast of L'Anse, Baraga County, MI	Neil J. Dougherty, superintendent 1894—produced 2,000 cu. ft. 1895—produced 23,000 cu. ft. 1896—produced 21,000 cu. ft. 1897—market dull, extracted 21,000 cu. ft. but shipped only 7,000 cu. ft.
Laughing Whitefish Stone Quarry (Craig and Wagner Quarry)		George Craig Henry D. Smith George Wagner 1871 additional investors: W. E. Deakman, Charles Moesinger, Charles Weunter	Marquette, MI Chicago, IL	1870–71	S25, 26, T48N R22W 30 ft. from mouth of Laughing Whitefish River at Rockport and one mile inland, Laughing Whitefish Point, Alger County, MI	1871—built settlement of Rockport, constructed tramway and dock, opened quarry and extracted stone by blasting By September 1873—operations suspended at a loss; rock not uniformly good quality
Marquette and Lake Superior Co.				1839	S35 T48N R25W Mount Mesnard, Marquette County, MI	1889—quarry opened and under development

Company Individual	Capital Stock	Stockholders	Main Office	Incorporation Date	Quarry Locations	Comments
Marquette Brownstone Co.	$500,000	Sidney Adams Thomas B. Brooks William Burt S. P. Ely Alfred Green John H. Jacobs Henry R. Mather Raphael Pumpelly F. P. Wetmore Peter White	Marquette, MI	1872	S 26 T48N R25W South Marquette, Marquette County, MI	John H. Jacobs oversaw operation 1878—quarry leased to Watson and Palmer
Michigan Red Stone Co.		Chas. C. Bloomfield Lawrence H. Field Addison B. Robinson Eugene J. Weeks Thomas A. Wilson	Jackson, MI	1891	S2 T52N R33W? Portage Entry, Houghton County, MI	1891—opened Peter A. Pickel, president, J. W. Wyckoff, superintendent Leased site from Martin Messner 1893—extracted 104,267 cu. ft.
R. B. Moss/Anthony Ping Quarry		New York capitalists		by 1889	S26 T48N R25W South Marquette, Marquette County, MI	1889—channeled and test pitted to determine prospects for quarry
Newport and Lake Superior Brownstone Co.	$100,000	Richard Blake Frederick W. Gillett John H. Gillett	Marquette, MI	1888	Lots 2, 3, 4, 5 S30 T52N R27W mouth of Salmon Trout River, Marquette County, MI	Robert Wagner, construction supervisor 1888—worker's houses, blacksmith shop, store, boardinghouse, stable built; fifty workers employed 1902—ceased business
Pike's Quarry		Robinson D. Pike	Bayfield, WI	1883	S33 T50N R4W Van Tassell's Point, 3.5 miles south of Bayfield, Bayfield County, MI	1883 to at least 1897—operated continuously 1889—produced 102,000 cu. ft. 1890—produced 77,473 cu. ft. 1892—produced 105,000 cu. ft.

Portage Entry Quarries Co.	$1,000,000	Henry Furst Ernst Heldmaier Edwin T. Malone Peter W. Neu Peter A. Pickel John Thomlinson	Chicago, IL	1893	S18, 19, T53N R32W NE1/2 Lot 7, S13 T53N R33W S6 T56N R31W Portage Entry, Houghton County, MI S2 T52N R33W Baraga County, MI S26 T48N R25W South Marquette, Marquette County, MI S 19, 20 T50N R8W Flag River near Port Wing, Bayfield County, WI	Consolidation of Furst, Neu and Co. and Portage Entry Red Stone Co. J. W. Wyckoff, manager until 1909 1895—shipped 219,525 cu. ft. 1901—George Froney, superintendent
Portage Entry Red Stone Co. (Portage Entry Red Sandstone Co.)		Malone Brothers	Cleveland, OH	by 1887	SE1/4 S18 T53N R32W Portage Entry, Houghton County, MI	Property owned by Earl Edgerton of L'Anse P. B. Parker, superintendent 1892—shipped 197,100 cu. ft. of block stone and employed seventy-five workers 1893—consolidated with Furst, Neu and Co. as Portage Entry Quarries Co.
Portage Entry Sandstone Co.	$1,000,000	Joseph Croze Nathan F. Leopold Philip B. Parker Prosper Roberts Norbert Sarazin	Cleveland, OH	1891	S31 T54N R32W Houghton County, MI	Shipped stone by scow to Marquette and then by rail to other points 1893—consolidated with Furst, Neu and Co. as Portage Entry Quarries Co.
Port Wing Quarry Co.		Miller Brothers Johnson	Duluth, MN	1895	S19, 20? T50N R8W 1.5 miles west of Port Wing, Bayfield County, WI	Top grade stone similar to Marquette "raindrop" 1895–97 shipped over 100,000 cu. ft. 1898—still operating
Powell's Point Quarry		Schoolcraft Iron Co. Bay Furnace Co.		1868–70	S26, 27 T27N R19W Powell's Point, Alger County, MI	Stone used in the construction of two neighboring blast furnaces
Powell's Point Quarry				1920s–1930s	S26, 27 T47N R19W Powell's Point, Alger County, MI	Stone used in the construction of the Lincoln School and Sacred Heart Convent, Munising

Company Individual	Capital Stock	Stockholders	Main Office	Incorporation Date	Quarry Locations	Comments
Prentice Brownstone Co.	$1,250,000	George H. Barr Edwin Ellis Cassius M. Hamilton Frederick Prentice Eugene A.Shores	Ashland, WI	1888 1891–93	S27 T49N R4W three miles south of Washburn at Houghton Point, Bayfield County, WI SE 1/4 S13, T51N R3W Hermit Island, Ashland County, WI	Built the town Houghton for workers 1889—purchased Cook and Hyde's real and personal property on island and on mainland 1889—shipped 383,887 cu. ft. 1890—shipped 623,334 cu. ft. Stone shipped by rail or by ship 1893—bankrupt 1897—Prentice and the Excelsior Brownstone Company relinquished title to all of Hermit Island to the estate of Elias Drake 1902—sold to W. G. Maginnes of New York City
Rock River Brownstone Co.	$500,000	George Barnes Willard I. Brotherton Charles Johnston Francis M. Moore Charles H. Shaffer Edward M. Watson	Marquette, MI	1889	S14, 15, 22, 23 T48N R21W east of Rock River between Lake Superior and DSS&A Railroad, Alger County, MI	1889—Charles Johnston located top-grade stone 1890–91—cleared site, built dock, railroad siding, houses, and a mill, and produced stone 1893—bankrupt 1902—ceased business
Herman Ruonawaara Quarry		Herman Ruonawaara			Across river from Craig Portage Entry, Houghton County, MI	
Schureman Quarry (Thomas A. Trevethan Quarry)		Israel Schureman Jacob Schureman	Chicago, IL	1889	Thomas A. Trevethan Farm on DSS&A railroad, six miles east of Houghton, Houghton County, MI	Schuremans were marble dealers in Chicago Reduction in freight rates by railroads permitted development of this landlocked quarry 1885—quarry worked in a limited way 1894—operated extensive rubble stone business By 1889 sold 80 acres to Lake

Strong, French and Co.			Chicago, IL	1870	Lots 1 and 2 SE1/4 SW1/4 S4 T50N R3W Basswood Island, Ashland County, WI	1873—closed operations
Superior Brownstone Co.	$35,000	Freeborn C. Bailey George H. Barr James H. Rogers	Ashland, WI	1891	SE1/4 SW1/2 S4 T50N R3W south end of Basswood Island, Ashland County, WI	Operated 1891–93 F. C. Bailey, superintendent W. H. Singer, president and general manager Stone sold from company's docks at Ashland, West Superior, and Duluth or shipped inland by rail
Superior Natural Redstone Co.		Lawrence J. DesRochers Raymond P. DesRochers Robert DesRochers		1959?	S2 T52N R33W Baraga County, MI	
Superior Red Sandstone Co.	$50,000	Ernest Bollman J. B. Cooper J. H. Rice (Pine?)		1899	S10 T52N R33W Baraga County, MI	1965?—ceased operations
John Thoney Quarry				1883	S35 T50N R26W .75 mile north of Little Garlic (Garlick) River at Thoney's Point, Marquette County, MI	1883—stone extracted in stripping shipped to Grand Marais, MI, for use in riprapping cribs at harbor of refuge Stone inappropriate for fancy ornamental work
Torch Lake Sandstone Quarry		Calumet and Hecla Mining Co.		By 1880	S1 T55N R33W west of Lake Linden near its rail line between the mine and stamping mill, Houghton County, MI	1882—submitted specimens of stone to Boston for pressure tests Stone shipped by rail to Calumet and Lake Linden for Calumet and Hecla Mining Company's industrial buildings

Company Individual	Capital Stock	Stockholders	Main Office	Incorporation Date	Quarry Locations	Comments
Traverse Bay Red Stone Co.		Charles Hebard	Pequaming, MI	1894–95	S6 T56N R31W at headwaters of Trap Rock River north of Lake Linden, Keweenaw County, MI	1895—constructed eight miles of railroad from the quarry to Traverse Bay, built dock, and shipped 6,500 cu. ft. 1896—shipped 20,000 cu. ft. and ceased operations Produced only a small quantity of building stone for local use Later, the Portage Entry Quarries Co. operated the quarry under lease, and J. W. Wyckoff was general manager
Washburn Stone Co.		C. W. Babcock W. H. Smith	Kasota, MN	1885	S27 T49N R4W Houghton, 15 miles north of Ashland, Bayfield County, WI	1885—quarry opened 1889—produced 105,000 cu. ft. 1890—produced 125,438 cu. ft. 1892—produced 185,000 cu. ft. 1897—still operating
Watson and Palmer		Edward M. Watson E. B. Palmer	Duluth, MN	1878	S26 T48N R25W South Marquette, Marquette County, MI	1878—leased site from Marquette Brownstone Co. and operated it until at least 1884 John H. Jacobs oversaw operation until 1881
Wieland Bros. Quarry			Duluth, MN	1905	S? T48N R4W one mile from Washburn at Anderson Falls, Bayfield County, WI	
Wolf, Jacobs and Co.		Peter Wolf John H. Jacobs	Chicago, IL	1883	S26 T48N R25W South Marquette, Marquette County, MI Lot 1, S19 T53N R32W Portage Entry, Houghton County, MI	1887—succeeded by Furst, Jacobs and Co. 1883—leased lot in S19 T53N R32W from Lake Superior Brownstone Co.
Peter Wolf and Son Co.		Peter Wolf M. W. Wolf	Chicago, IL	1869	S26 T48N R25W Pendill farm site South Marquette, Marquette County, MI	1869—purchased from George Craig 1870—John H. Jacobs oversaw operation 1872—sold to Marquette Brownstone Co.

APPENDIX 2

STOCKHOLDERS/INVESTORS IN SIGNIFICANT QUARRY COMPANIES IN THE JACOBSVILLE FORMATION AND THE BAYFIELD GROUP

Name		Residence	Company
Adams	Sidney	Marquette, MI	Marquette Brownstone Co.
Albrecht	A. A.	Detroit, MI	Detroit and Marquette Brownstone Co.
Amberg	W. A.	Chicago, IL	L'Anse Brownstone Co.
Babcock	C. W.	Kasota, MN	Washburn Stone Co.
Badgley	T. V.	Superior, WI	Iron River Red Sandstone Co.
Bailey	Freeborn C.	Ashland, WI	Superior Brownstone Co.
Barnes	George	Marquette, MI	Butler Brownstone Co.
Barnes	George	Marquette, MI	Lake Superior Red Stone Co.
Barnes	George	Marquette, MI	Rock River Brownstone Co.
Barr	George H.	Ashland, WI	Prentice Brownstone Co.
Barr	George H.	Ashland, WI	Superior Brownstone Co.
Bending	F.	Marquette, MI	Building Stone and Mineral Exploring Co.
Berry	Joseph H.	Detroit, MI	Detroit and Marquette Brownstone Co.
Berry	Thomas	Detroit, MI	Detroit and Marquette Brownstone Co.
Bissell	John H.	Detroit, MI	Detroit and Marquette Brownstone Co.
Blake	Richard	Marquette, MI	Newport and Lake Superior Brownstone Co.
Bloomfield	Chas. C.	Jackson, MI	Michigan Red Stone Co.
Bodenschatz	G. A.	Chicago, IL	Ashland Brown Stone Co.
Bodenschatz	J. G.	Chicago, IL	Ashland Brown Stone Co.
Bollman	Ernest		Superior Red Sandstone Co.
Breckenridge	John	Kentucky	Breckenridge Quarry
Breitung	Edward	Marquette, MI	L'Anse Brown Stone Co.
Brooks	Thomas B.	Marquette, MI	Marquette Brownstone Co.
Brooks	Walter C.	Superior, WI	Bass Island Brown Stone Co.
Brotherton	Willard I.	Bay City, MI	Rock River Brownstone Co.
Brown	E. H.	Chicago, IL	Ashland Brown Stone Co.

Burt	A. Judson	Marquette, MI	Burt Free Stone Co.
Burt	Hiram A.	Marquette, MI	Burt Free Stone Co.
Burt	John	Marquette, MI	Burt Free Stone Co.
Burt	William	Marquette, MI	Burt Free Stone Co.
Burt	William	Marquette, MI	Burt Free Stone Co.
Burt	William A.	Marquette, MI	Marquette Brownstone Co.
Butler	Thomas	Au Train, MI	Butler Brownstone Co.
Call	C. H.	Marquette, MI	Building Stone and Mineral Exploring Co.
Call	C. H.	Marquette, MI	Lake Superior Red Stone Co.
Candler	I.	Detroit, MI	Detroit and Marquette Brownstone Co.
Candler	W. R.	Detroit, MI	Detroit and Marquette Brownstone Co.
Candler	H. W.	Detroit, MI	Detroit and Marquette Brownstone Co.
Carter	Howard M.	Chicago, IL	Furst, Neu Co.
Case	D. L.	Marquette, MI	Keweenaw Redstone Co.
Cheney	A. J.	Kenosha, WI	Arcadian Brownstone Co.
Cochran	V. B.	Marquette, MI	Building Stone and Mineral Exploring Co.
Cochran	Varnum B.	Marquette, MI	Lake Superior Brownstone Co.
Cook	Thomas D.	Milwaukee, WI Minneapolis, MN	Cook and Hyde
Cooper	J. B.		Superior Red Sandstone Co.
Craig	George	Marquette, MI	Laughing Whitefish Stone Quarry
Craig	Thomas	Marquette, MI	Laughing Whitefish Stone Quarry
Crebassa	Peter	L'Anse, MI	
Crowley	John H.	Duluth, MN	Duluth Brownstone Co.
Croze	Joseph	Houghton, MI	Portage Entry Sandstone Co.
Dalliba	William S.	Marquette, MI	Lake Superior Brownstone Co.
David	M. E.	Marquette, MI	Detroit and Marquette Brownstone Co.
David	M. W.	Marquette, MI	Detroit and Marquette Brownstone Co.
David	Myron E.	Marquette, MI	Detroit and Marquette Brownstone Co.
David	O. I.	Edwards, NY	Detroit and Marquette Brownstone Co.
Deakman	Wm. F.	Chicago, IL	Laughing Whitefish Stone Quarry
Denison	Francis W.	Bayfield, WI	Bayfield Brownstone Development Co.
DesRochers	Lawrence J.	Arnheim, MI	Superior Natural Redstone Co.
DesRochers	Raymond P.	Arnheim, MI	Superior Natural Redstone Co.
DesRochers	Robert	Arnheim, MI	Superior Natural Redstone Co.
Ellis	Edwin	Ashland, WI	Excelsior Brownstone Co.
Ellis	Edwin	Ashland, WI	Prentice Brownstone Co.

Ely	S. P.	Marquette MI	Marquette Brownstone Co.
Field	Lawrence	Jackson, MI	Michigan Red Stone Co.
Fifield	S. S.	Ashland, WI	Excelsior Brownstone Co.
Findeisen	Carl	Chicago, IL	Kerber-Jacobs Redstone Co.
French	Edwin C.	Milwaukee, WI Chicago, IL	Bass Island Brown Stone Co.
Furst	George C.	Chicago, IL	Furst, Jacobs and Co.
Furst	Henry	Chicago, IL	Portage Entry Quarries Co.
Furst	Henry	Chicago, IL	Furst, Neu Co.
Furst	Henry	Chicago, IL	Furst, Jacobs and Co.
Furst	Henry	Chicago, IL	Furst, Neu and Co.
Garnett, Jr.	William	Chicago, IL	Furst, Neu Co.
Gillett	Frederick W.	Marquette, MI	Newport and Lake Superior Brownstone Co.
Gillett	John H.	Marquette, MI	Newport and Lake Superior Brownstone Co.
Goddard	F. E.	Ashland, WI	Excelsior Brownstone Co.
Goodale	S. W.	Marquette, MI	Kerber-Jacobs Redstone Co.
Green	Alfred	Marquette, MI	Marquette Brownstone Co.
Hamilton	Cassius M.	Ashland, WI	Prentice Browntone Co.
Hebard	Charles	Pequaming, MI	Traverse Bay Red Stone Co.
Heldmaier	Ernst	Chicago, IL	Furst, Jacobs and Co.
Heldmaier	Ernst	Chicago, IL	Furst, Neu Co.
Heldmaier	Ernst	Chicago, IL	Portage Entry Quarries Co.
Hill	W. S.	Marquette, MI	Building Stone and Mineral Exploring Co.
Hollands	I. G.	Detroit, MI	Detroit and Marquette Brownstone Co.
Horton	J. H	Superior, WI	Iron River Brownstone Co.
Humphrey	H. Hubert	Detroit, MI	Detroit and Marquette Brownstone Co.
Hurley	Timothy T.	Marquette, MI	L'Anse Brown Stone Co.
Hyde	Edwin	Milwaukee, WI Minneapolis, MN	Cook and Hyde
Ingalls	Edmund	Duluth, MN	Ingalls Quarry
Jacobs	J. H.	Marquette, MI	Building Stone and Mineral Exploring Co.
Jacobs	J. H.	Marquette, MI	Kerber-Jacobs Redstone Co.
Jacobs	John H.	Marquette, MI	Furst, Jacobs and Co.
Jacobs	John H.	Marquette, MI	Wolf and Jacobs Co.
Jacobs	J. Henry	Marquette, MI	Marquette Brownstone Co.
Johnson	Otis W.	Racine, WI	Excelsior Redstone Co.
Johnson		Duluth, MN	Port Wing Quarry Co.

Johnston	Charles	Rock River, MI	Butler Brownstone Co.
Johnston	Charles	Rock River, MI	Rock River Brownstone Co.
Kennedy	D. S.	Ashland, WI	Ashland Brown Stone Co.
Kidder	Alfred	Detroit, MI	Detroit and Marquette Brownstone Co.
Kidder	Alfred	Marquette, MI	Lake Superior Brownstone Co.
Knight	John H.	Ashland, WI	Ashland Brown Stone Co.
Knight	William	Ashland, WI	Ashland Brown Stone Co.
Lane	A. L.	Detroit, MI	Detroit and Marquette Brownstone Co.
Lawrence	Leonard	Detroit, MI	Detroit and Marquette Brownstone Co.
Lazier	Frank B.	Duluth, MN	Duluth Brownstone Co.
Leach	Cambreton	Superior, WI	Bass Island Brown Stone Co.
Lee	George P.	Milwaukee, WI Chicago, IL	Bass Island Brown Stone Co.
Leopold	Nathan F.	Chicago, IL	Portage Entry Sandstone Co.
Little	Charles H.	Detroit, MI	Detroit and Marquette Brownstone Co.
Luling	Theo. W.	Marquette, MI	Keweenaw Redstone Co.
Malone	Brothers	Cleveland, OH	Portage Entry Red Stone Co.
Malone	Edwin T.	Chicago, IL	Portage Entry Quarries Co.
Mapes	Egbert J.	Marquette, MI	Lake Superior Brownstone Co.
Mather	Henry R.	Marquette, MI	Marquette Brownstone Co.
Maxwell	W. S.	Kenosha, WI	Arcadian Brownstone Co.
McDougall	Alexander	Duluth, MN	Duluth Brownstone Co.
Miller		Duluth, MN	Flag River Brown Stone Co.
Miller	Brothers	Duluth, MN	Port Wing Quarry Co.
Mitchell	John	Marquette, MI	Building Stone and Mineral Exploring Co.
Mitchell	John	Marquette, MI	Kerber-Jacobs Redstone Co.
Moesinger	Charles	Chicago, IL	Laughing Whitefish Stone Quarry
Moore	F. M.	Marquette, MI	Butler Brownstone Co.
Moore	Francis M.	Marquette, MI	L'Anse Brown Stone Co.
Moore	Francis M.	Marquette, MI	Rock River Brownstone Co.
Morse	Jay C.	Marquette, MI	Lake Superior Brownstone Co.
Neu	Anna M.	Chicago, IL	Furst, Neu Co.
Neu	Peter W.	Chicago, IL	Furst, Jacobs and Co.
Neu	Peter W.	Chicago, IL	Furst, Neu and Co.
New	Peter W.	Chicago, IL	Portage Entry Quarries Co.
O'Brien	William	Chicago, IL	L'Anse Brownstone Co.
Omeis			Cranberry River Quarry

Packard	B. S.	Marquette, MI	Building Stone and Mineral Exploring Co.
Packard	B. S.	Marquette, MI	Kerber-Jacobs Redstone Co.
Palmer	E. B.	Duluth, MN	Watson and Palmer
Parker	Philip B.	Houghton, MI	Portage Entry Sandstone Co.
Pickel	Peter A.	Chicago, IL	Furst, Jacobs and Co.
Pickel	Peter A.	Chicago, IL	Furst, Neu Co.
Pickel	Peter A.	Chicago, IL	Portage Entry Quarries Co.
Pike	Robinson D.	Bayfield, WI	Bayfield Brownstone Development Co.
Pike	Robinson D.	Bayfield, WI	Pike's Quarry
Pine	Paul	Marquette, MI	L'Anse Brown Stone Co.
Pope	Willard S.	Detroit, MI	Detroit and Marquette Brownstone Co.
Powell	D. W.	Marquette, MI	Building Stone and Mineral Exploring Co.
Powell	D. W.	Marquette, MI	Kerber-Jacobs Redstone Co.
Powell	D. W.	Marquette, MI	Lake Superior Red Stone Co.
Prentice	Frederick	Ashland, WI New York, NY	Prentice Brownstone Co.
Pumpelly	Raphael	Cambridge, MA	Marquette Brownstone Co.
Quayle	George H.	Bayfield, WI	Bayfield Brownstone Development Co.
Quinby	F. H.		Cranberry River Quarry
Rankin	Ernest	Marquette, MI	Lake Superior Red Stone Co.
Rice (Pine?)	J. H.		Superior Red Sandstone Co.
Ritchie	James S.	Superior, WI	Bass Island Brown Stone Co.
Roberts	Prosper	Lake Linden, MI	Portage Entry Sandstone Co.
Roberts	S. B.	Superior, WI	Iron River Red Sandstone Co.
Robinson	Addison B.	Jackson, MI	Michigan Red Stone Co.
Robson	Frank E.	Detroit, MI	Excelsior Redstone Co.
Rogers	James H.	Ashland, WI	Superior Brownstone Co.
Ruonawaara	Herman		Herman Ruonawarra Quarry
Sarazin	Norbert	Lake Linden, MI	Portage Entry Sandstone Co.
Schaffer	C. H.	Marquette, MI	Butler Brownstone Co.
Schureman	Israel	Chicago, IL	Schureman Quarry (Trevethan Quarry)
Schureman	Jacob	Chicago, IL	Schureman Quarry (Trevethan Quarry)
Seager	James	Hancock, MI	Hancock Sandstone Land Co.
Seager	James B.	Hancock, MI	Excelsior Redstone Co.
Seager	James H.	Hancock, MI	Excelsior Redstone Co.
Shaffer	Charles H.	Marquette, MI	Rock River Brownstone Co.
Shores	Eugene A.	Ashland, WI	Prentice Brownstone Co.

Smith	Gad	Marquette, MI	Building Stone and Mineral Exploring Co.
Smith	Gad	Marquette, MI	Lake Superior Red Stone Co.
Smith	Henry D.	Marquette, MI	Laughing Whitefish Stone Quarry
Smith	I. C.	Detroit, MI	Detroit and Marquette Brownstone Co.
Smith	Samuel L.	Detroit, MI	Excelsior Redstone Co.
Smith	W. H.	Kasota, MN	Washburn Stone Co.
Stafford	Henry	Marquette, MI	Lake Superior Brownstone Co.
Strong	Robert H.	Milwaukee, WI Chicago, IL	Bass Island Brown Stone Co.
Sweet	Alanson	Milwaukee, WI	Bass Island Brown Stone Co.
Thomlinson	John	Chicago, IL	Portage Entry Quarries Co.
Thoney	John	Marquette, MI	John Thoney Quarry
Tipton	R. I.	Superior, WI	Iron River Brownstone Co.
Titus	A. C.	Superior, WI	Iron River Brownstone Co.
Titus	Albert. C.	Superior, WI	Iron River Red Sandstone Co.
Tobias	L. C.	Ashland, WI	Excelsior Brownstone Co.
Tower	E. H.	Marquette, MI	Kerber-Jacobs Redstone Co.
Wagner	George	Marquette, MI	Laughing Whitefish Stone Quarry
Watson	Edward M	Marquette, MI	Rock River Brownstone Co.
Watson	Edward M.	Duluth, MN	Watson and Palmer
Weeks	Eugene J.	Jackson, MI	Michigan Red Stone Co.
Wells	Daniel	Milwaukee, WI Chicago, IL	Bass Island Brown Stone Co.
Wells	Daniel	Milwaukee, WI	Bass Island Brown Stone Co.
Wetmore	F. P.	Marquette, MI	Marquette Brownstone Co.
Weunter	Charles	Chicago, IL	Laughing Whitefish Stone Quarry
White	Peter	Marquette, MI	Marquette Brownstone Co.
Wieland	Brothers	Duluth, MN	Wieland Brothers Quarry
Wilson	Thomas A.	Jackson, MI	Michigan Red Stone Co.
Wolf	Peter	Chicago, IL	Wolf and Jacobs Co.
Wolff	M. W.	Chicago, IL	Peter Wolf and Son Co.
Wolff	Peter	Chicago, IL	Peter Wolf and Son Co.
Wood	Ira C.	Chicago, IL	Furst, Neu Co.

APPENDIX 3

SANDSTONE BUILDINGS IN THE LAKE SUPERIOR REGION

Note: Building names with asterisks indicate those discussed in the text.

Michigan

Alger County

Au Train

1890
Alger County Jail and Sheriff's Residence (destroyed 1900)
Demetrius Frederick Charlton, architect
Rock River Brownstone Company

Munising

1868
Schoolcraft Furnace*
Munising Street
Grand Island quarry

1895–96
First National Bank of Alger County
Ferguson Brothers, contractors
100 West Munising Avenue (Michigan 28)

1901–02
Alger County Courthouse (destroyed 1978)
Charlton and [R. William] Gilbert, architects
Northern Construction Company, contractors
Bounded by Jewell, Park, and Court Streets

c. 1902
Alger General Agency/Detroit and Northern Savings Bank
200 block of Elm Avenue
Brownstone and white stone

c. 1910
Sacred Heart School (Jerico House Apartments)
408 Elm Avenue

c. 1920
Sacred Heart Convent (destroyed)
Elm Avenue at East Jewell Street
Powell's Point quarry

1922
Lincoln School
Demetrius Frederick Charlton, architect
Superior Street
Powell's Point quarry

Onota

1869–70
Bay Furnace*
Just east of Christmas, northeast of Michigan 28 in Hiawatha National Forest (Lot 2, Section 29 of T47N R19W)
Powell's Point quarry

Baraga County

Assinins

1873, 1875, 1881
Church (destroyed), Rectory (destroyed), and Convent and Orphanage of the Catholic Mission (Keweenaw Bay Tribal Community)
US 41

L'Anse

1874
First Methodist Church
North Main Street
L'Anse Brownstone Company

1894
Sacred Heart Church★
Demetrius Frederick Charlton, architect
16 South Sixth Street
L'Anse Brownstone Company

Chippewa County

Sault Sainte Marie

1877, 1904, 1988–89
Chippewa County Courthouse
William Scott, architect
1904 addition, R. C. Sweatt, architect
1988–89 restoration, Lincoln A. Poley, architect
Bounded by Maple Street, Bingham Avenue, Spruce Street, and Court Street
Marquette brownstone trim

1887
Sault Saint Marie National Bank (destroyed 1896)
Northeast corner Portage Avenue and Ashmun Street
Brick with sandstone foundation and trim

1893–94
Central Methodist Episcopal Church
Dillon P. Clark, architect
111 East Spruce Street (East Spruce and Court Streets)
Red and gray variegated canal stone

1895
Lipsett Block
Ashmun Street
Portage Entry

Lipsett House
315 Spruce Street
Portage Entry

Soo Line Depot
Portage

1896–1902
Michigan Lake Superior Hydroelectric Power Plant (Edison Sault Power Plant)★
Hans A. E. von Schon, engineer; James Calloway Teague, architect for superstructure
725 East Portage Avenue, over the power canal and on the Saint Marys River, off East Portage Avenue, Johnston Street, and Union
Canal stone

1897
East Spruce Street School (Garfield School)
Spruce Street between Canal and Johnson
Canal stone?

c. 1898
Sault Sainte Marie News Building
115 Ashmun Street
Portage Entry

c. 1900
The Antlers
804 East Portage
Canal rock

Comb Building
Portage Avenue
Canal rock

Commercial Building
226 Ashmun
Canal rock

D. K. Moses and Company Dry Goods Building
500 block Ashmun
Canal rock

Everett Block
519 Ashmun Street
Red and gray variegated canal rock

Hewitt Grocery (Soo Supply)
224 East Portage Avenue
First story front smooth-cut red sandstone

House
303 Armory Street
Red sandstone

Vanderhook Furniture
539 Ashmun Street
Red sandstone front, white rubble stone side walls

Water Tower
Southeast corner Ryan Street and Easterday Avenue
Canal rock

1902
First United Presbyterian Church
Edward Demar, architect
309 Lyon
Portage Entry

McTavish Building (Murray Hill Hotel) (destroyed 1940s)
106-116 Maple Street
Canal rock

1902–3
Saint James Episcopal Church
James Calloway Teague, architect
533 Bingham Avenue

1904
Newton Block
Edward Demar, architect
506 Ashmun Street
Sandstone trim

Gogebic County

Bessemer

1888, 1915
Gogebic County Courthouse
John Scott and Company, architects
Wahlman and Grip, contractors
1915 rebuilding and enlargement, Charlton and Kuenzli, architects
Portage Entry

Ironwood

1896
Church of the Transfiguration
Demetrius Frederick Charlton, architect
East Aurora and South Marquette

1909
Ironwood Station of the Chicago and Northwestern Railroad Company (Ironwood Area Historical Museum)
Charles S. Frost, architect
Off Frederick Street, between Suffolk and Lowell Streets
Portage Entry

Houghton County

Calumet

1880s, 1899–1900
Vertin Brothers Department Store*
1899 remodeling, Charles W. Maass, architect
Paul P. Roehm, stone mason
220 Sixth Street
Portage Entry

1886, 1899
First National Bank (destroyed)
J. B. Sweatt, architect and builder
1899 alterations and addition, Charles K. Shand, architect
Paul P. Roehm, stone mason
Fifth Street
Portage Entry

1895
Agnitz Block
Henry Key, contractor
427 Fifth Street
Portage Entry

Commercial Building
425 Fifth Street
Portage Entry

1896
Saint Mary's Italian Church (The Assumption of the Blessed Virgin Church)*
Martin Brothers, contractor

South side of Portland Street between Ninth and Tenth Streets
Portage Entry

1897
Michael Nampaa Building

1898
Blau Block (destroyed 1981)
813 Elm Street

Bernard Bracco Building
Charles K. Shand, architect
Elm Street

Joseph Hermann Building
Charles W. Maass, architect
106 Sixth Street
Portage Entry

Kinsman Block
Charles W. Maass, architect
Paul P. Roehm, masonry contractor
101 Sixth Street
Portage Entry

Edward Ryan Block
Charles K. Shand, architect
L. E. Chausse, contractor
305-307 Sixth Street
Portage Entry

Joseph Suino Block (destroyed)
Scott, architect
Corner of Sixth and Portland Streets
Portage Entry

1898–99
Red Jacket Fire Station★
Charles K. Shand, architect
Procissi and Company, stone masonry
Sixth Street, just south of the southwest corner of Elm Street
Portage Entry

1899–1900
Red Jacket Town Hall and Opera House (Calumet Village Hall and Calumet Theater)★
Charles K. Shand, architect
Paul P. Roehm, stone masonry
340 Sixth Street
Portage Entry

1900
Ernest Bollman Block (destroyed)
Charles K. Shand, architect
Paul P. Roehm, masonry contractor
Southwest corner of Oak and Eighth Streets

Gately-Wiggins Block (destroyed)
[Clarence L.] Cowles and [George] Eastman, architects
Sixth Street

N. P. Swanson Block (destroyed)
Carl E. Nystrom, architect
Elm Street
Portage Entry

1900–1901
William Craze Block (planned)
Eighth Street near Oak Street

1902
V. Coppo Block
213 Sixth Street
Portage Entry

1902–4
Saint Joseph's Austrian (Slovenian) Church (Saint Paul the Apostle Church)★
Charles K. Shand[?], architect
Paul P. Roehm, contractor
Northwest corner of Eighth and Oak Streets
Portage Entry

1906
First National Bank
Southeast corner of Sixth and Oak Streets

Calumet Township

1880–83
Calumet and Hecla Water Works (destroyed)
Calumet Waterworks Road at Lake Superior

1895
Swedish Methodist Episcopal Church
Bajari and Ulseth, contractor
3907 Sixth Street
Portage Entry

1897
YMCA Building (Elk's Temple)
Southeast corner of Sixth Street and Wedge
Paul P. Roehm, stone contractor
Portage Entry

1899–1901
Sainte Anne's Roman Catholic Church (Keweenaw Heritage Center)*
Charlton, Gilbert and [Edward] Demar, architects
Prendergast and Clarkson, contractor
Southeast corner of Fifth and Scott Streets
Portage Entry

1900
Calumet Brewing Company (in ruins)
Tamarack Water Works Road near intersection with Dextron Road
Portage Entry

Franklin Township

1894–95
Seth D. North and Son Store (destroyed)
Edward Demar of Demar and Lovejoy, architect
E. E. Grip and Company, contractor
Portage Entry

1895–97
Quincy Mining Company Office (Pay Office)
Robert C. Walsh, architect
West side of US 41 at Quincy Mine Location
Portage Entry

1900
Blacksmith and Drill Shop, Quincy Mining Company
East side of US 41 at Quincy Mine Location
Variegated Portage Entry

Hancock

1890s
Double House
534-536 Quincy Street (US 41)
Portage Entry

1895
Andrew Johnson Block (execution unknown)
Reservation Street

1898
Edward Lieblein Building
Hancock Street near railroad track
Portage Entry

1898–99
Hancock Town Hall and Fire Hall (Hancock City Hall)*
Charlton, Gilbert and Demar, architects
1984–85, rehabilitation and addition, Francis J. Rutz, Hitch, Inc.
399 Quincy Street (US 41)
Portage Entry

1898–1900
Finnish Lutheran College and Seminary (Suomi College)*
C. Archibald Pearce, architect
Quincy Street (US 41) at Dakota Street
Portage Entry

Hancock Hotel (planned, not executed)
Charlton, Gilbert and Demar, architects
Corner of Reservation Street, opposite the Jacob Baer House and on the site of the old fire station

1899
S. F. Prince Building (planned)
Quincy Street (US 41)

1899–1900
C. D. Hanchett Building (demolished?)
Charlton, Gilbert and Demar, architects
Southwest corner of Quincy (US 41) and Reservation Streets

1900
House
907 Quincy Street (US 41)
Portage Entry

Kauth Block
H. T. Liebert, architect
John H. Foster and Son, contractor
Quincy Street (US 41)

Charles A. Wright Block
Charlton, Gilbert and Demar, architects
Prendergast and Clarkson, contractors
100-102 Quincy (US 41) Street
Portage Entry

c. 1900
Commercial Block
325 Quincy Street (US 41)

Commercial Building
632 Quincy Street (US 41)

1905
Israel and Rachel Epstein Building
116 Quincy Street (US 41)

1906
Scott Hotel
101 East Quincy Street

c. 1910
House
923 Summit

Houghton

1875
Adam Haas Brewery (destroyed 1968)
405 Shelden Avenue (US 41)

W. Miller Block
413 Shelden Avenue (US 41)

by 1883
Union School (destroyed)

1887–89
Michigan Mining School (Hubbell Hall/Science Hall) (destroyed 1968)*
John Scott and Company, architects
Wahlman and Grip, contractors
College Avenue (US 41)
Portage Entry red stone; Furst, Jacobs and Company, Portage Entry

1888–89
National Bank of Houghton*
Scott and Company, architects
Wahlman and Grip, contractors
600 Shelden Avenue (US 41)
Portage Entry

1892–93, 1918–25
Grace Methodist Church
William T. Pryor, architect
1990–91, addition and rehabilitation, Francis J. Rutz, Hitch, Inc.
201 Isle Royale Street
Portage Entry

1898–1902, 1928, 1956
Saint Ignatius Loyola Church
Erhard Brielmaier and Sons, architect
Fred E. King and Company, masonry contractor
703 East Houghton Avenue
Portage Entry

1899
James R. Dee Building
Henry L. Ottenheimer, architect
Paul F. P. Mueller, contractor
65 North Isle Royale Street
Portage Entry

Shelden-Dee Block*
Henry L. Ottenheimer, architect
Paul F. P. Mueller, contractor
512-524 Shelden Avenue (US 41)
Portage Entry

1899–1900
F. W. Kroll Block
Charlton, Gilbert and Demar, architects
Wilson and Sampson, contractors
606 Shelden Avenue (US 41)

c. 1900
Commercial Block
122-126 Shelden Avenue (US 41)
Portage Entry

Hall Building
320-322 Shelden Avenue (US 41)
Portage Entry

1901
Citizens' National Bank
C. Archibald Pearce, architect
Shelden Avenue (US 41)

1903
Hoar Mausoleum
Forest Hill Cemetery
Portage Entry

Houghton Station of the Duluth, South Shore and Atlantic Railway (Soo Line Railroad Company)
Portage Entry

c. 1905
Houghton Masonic Temple Building (City Centre)*
Charles W. Maass and Fred A. H. Maass, architects
616-618 Shelden Avenue (US 41)
Portage Entry

c. 1910
Foley Block
404-408 Shelden Avenue (US 41)
Portage Entry

Hubbell

1902–3
Saint Cecilia's School
1401 Euclid Street
Portage Entry

Lake Linden

1891
Hotel Linden (destroyed)
Portage Entry

1898–1900
First National Bank
346 Third Street (Michigan 26)
Portage Entry

1900, 1912
Eglise Saint Joseph (Saint Joseph Church)
C. Archibald Pearce, architect
1912 alterations, Demetrius Frederick Charlton, architect
701 Calumet Street (Michigan 26)
Portage Entry

1901–2
Lake Linden Village Hall and Fire Station
Charles K. Shand, architect
L. F. Ursin, contractor
401 Calumet Street (Michigan 26)
Portage Entry

Laurium

c. 1895
Commercial Building
317 Hecla Street
Portage Entry

c. 1895–96
Paul P. and Anna L. Roehm House*
Paul P. Roehm, builder
101 Willow Avenue
Portage Entry

1896–98
Sacred Heart Church (destroyed)
Louis Picket with Charles W. Maass, architects
Paul P. Roehm, stone masonry contractor
1383 Rockland Street
Portage Entry

1897
Dominick Quello Block
John Procissi, contractor
Hecla Street near the Laurium Village Hall
Portage Entry

1898
Johnson and Anna Lichty Vivian, Jr., House*
Demetrius Frederick Charlton, architect
Northeast corner of Pewabic and Third Streets
Portage Entry

1899
W. H. Faucett and Brothers Block
Paul P. Roehm, stone work contractor
Hecla Street (Michigan 26)

Manier and Hunt Hotel (destroyed)
Charles W. Maass, architect
Corner of Hecla Street (Michigan 26) and Torch Lake Road

John R. O'Neil Block (destroyed)
Charles K. Shand, architect
Donahue, stone contractor
Osceola and Third Streets
Portage Entry

Tinette Block
116 Osceola Street
Portage Entry

1899, 1912
Laurium Village Hall
1899 design and 1912 remodeling, Charles W. Maass, architect
310 Hecla Street (Michigan 26)
Portage Entry

1900
Peter Contralto Block
Cowles and Eastman, architects
201-203 Hecla Street (Michigan 26)
Portage Entry

D. Marta Block
323 Hecla Street (Michigan 26)
Portage Entry

Michael A. Richetta Block
c. 1900
Commercial Building
316-322 Hecla Street (Michigan 26)
Portage Entry

c. 1900–1905
Commercial Building
Northeast corner of Pewabic and First Streets
Portage Entry

1901
State Savings Bank (Great Lakes Mutual Insurance Company)
Carl E. Nystrom, architect
Southeast corner of Hecla (Michigan 26) and Third Streets
Portage Entry, columns of Saint Cloud granite

1904–7
Carlton House
401 Kearsarge Street
Portage Entry

Painesdale

1909–10, 1935
Jeffers High School (Painesdale High School)*
Alexander C. Eschweiler, architect
Goodell between Iroquois and Hulbert Avenue
Portage Entry

1902–3
Sarah Sargent Paine Memorial Library (destroyed 1964)*
Alexander C. Eschweiler, architect
Prendergast and Clarkson, contractor
Goodell
Portage Entry

Ripley

1914
Ripley School
Demetrius Frederick Charlton, architect
Michigan 26
Portage Entry

South Lake Linden

1886–87
Calumet and Hecla Smelting Works
1700 Duncan Avenue (Michigan 26)
Torch Lake quarry

South Range

1907–10
Kaleva Temple Block
Trimountain Avenue (Michigan 26) at Champion Avenue
Portage Entry

1935
South Range Community Building
Philip Verville, Hancock, contractor
Trimountain Avenue (Michigan 26) at Champion Avenue
Portage Entry

Marquette County

Ishpeming

1880–81
Engine House
Iron Cliffs Mining Company
Near the intersection of Michigan 28 and the Marquette, Houghton and Ontonagon Railroad switch

1885
Edward R. and Jennie Bigelow Hall House
112 Bluff

1890
Toutloff Building
E. E. Grip and Company, contractors
Portage Entry

1890–91
Ishpeming City Hall
Demetrius Frederick Charlton, architect
Sinclair and Outerson, contractors
100 East Division Street
Portage Entry

1891, 1902
Cliffs Cottage (William G. Mather Chalet)
Charlton and Gilbert, architects
1902 addition
282 Jaspar Street

1895–96
Ishpeming High School (destroyed c. 1981)
Charlton, Gilbert and Demar, architects
Corner of North and First

c. 1900
Anderson Block
Northeast corner of Main and Pearl
Portage Entry

Negaunee

c. 1885
Commercial Block
434 Iron
Marquette

1895
Water Works

1890s
Shea Block
300 block of Iron
Marquette

Marquette

1855
First Marquette County Jail (destroyed)
Charles Johnson, builder

1860s
John Burt House*
220 Craig Street

1868–70
Machine Shops and Roundhouse of the Marquette and Ontonagon Railroad Company*
Pendill farm site quarry (Peter Wolf quarry)

1870–73, 1891
First United Methodist Church
William Wyckoff, architect/builder; steeple design, Carl F. Struck, architect
1891 alterations, Demetrius Frederick Charlton
113 East Ridge Street
Marquette

1871
Grace Furnace of the Lake Superior Iron Company (destroyed)
Marquette

1871–73, 1885
Superior Building (First National Bank) (destroyed)*
Henry Lord Gay and Company, architect
Carl F. Struck, supervising architect
Alfred Green, contractor
1885 rebuilding, Hampson Gregory, architect/builder
Southwest corner of Front and Spring Streets
Wolf quarry

1872
Dock Office of the Marquette and Ontonagon Railroad
120 East Main Street
Marquette

1872–76
Hiram A. and Sarah Benedict Burt House (Sidney and Harriet Adams House)*
Carl F. Struck, architect
202 East Ridge Street
Burt Freestone Company

1874–75
High School (destroyed 1900)*
Carl F. Struck, architect
Ridge, Arch and Pine Streets
Marquette Brownstone Company

1874–77
First Opera House (destroyed)
Carl F. Struck, architect
Front Street between Washington and Main Streets

1874–76, 1887
Saint Paul's Episcopal Church*
Gordon W. Lloyd, architect, with Carl F. Struck, supervising architect
Growling, stone mason
Northeast corner of Ridge and High Streets
Marquette Brownstone Company
1887 addition, Morgan Memorial Chapel *
[Henry Ives] Cobb and [Charles S.] Frost, architects
Hampson Gregory, contractor
318 West High Street
Marquette variegated

1875
Andrew A. and Laura Greenough Ripka House*
Carl F. Struck, architect
430 East Arch Street
Marquette

1875–76
Richard P. Traverse House (destroyed)
Carl F. Struck, architect
135 Bluff Street
Marquette

1880
Daniel H. and Harriet Alford Merritt House*
Hampson Gregory, architect/builder
410 East Ridge Street
Marquette

c. 1880
Commercial Building
412 South Front Street

1880, 1935
Saint Peter's Roman Catholic Cathedral*
Henry G. Koch and Son, architects; Hampson Gregory, contractor; James Lawrence, masonry superintendent; Demetrius Frederick Charlton, interior woodwork design
1935 rebuilding, Edward A. Schilling, architect
311 West Baraga Street
Marquette

1881
James M. Pickands Warehouse
121 Baraga Street
Marquette

1883
Edward Fraser Block (Rosewood Inn)
211 South Front Street
Marquette

Martin Vierling Block
121 South Front Street
Marquette

1884–86
First Baptist Church (destroyed)
Henry W. Coddington, contractor
Southeast corner of Ridge and Front Streets
Burt Freestone Company

1885–89
Upper Peninsula Branch Prison and House of Correction*
William Scott and Company, architects, with Demetrius Fredrick Charlton, supervising architect
Wahlman and Grip, contractors
1960 US 41 South
Marquette, Portage Entry

1887
Edward and Mary Pauline Breitung House (planned, not executed)
William Scott and Company, architects
Ridge Street
Portage Entry

Breitung Mausoleum
John Scott and Company, architects
Furst and Neu, contractor
Park Cemetery
Portage Entry

Harlow Block*
Hampson Gregory, architect/builder
102 West Washington Street
Marquette "raindrop"

William Hicks Block
329 West Washington Street

First Timothy Nester Block (destroyed 1891)
Smith and Wilson, contractors
136 West Washington Street

1888–89
Campbell and Wilkinson Bank Block (destroyed)
John Scott and Company with Demetrius Frederick Charlton, architects
Northeast corner of Front and Washington Streets
Portage Entry

Herman B. Ely School (destroyed 1969)
Demetrius Frederick Charlton of Scott and Charlton, architect
Wilson and Moore, contractors
Bluff Street
Portage Entry

United States Custom House, Post Office, Court House, and Government Building (destroyed)*
M. E. Bell and Will A. Freret, architects
Smith and Wilson, contractors
John Lawrence, stone cutter
Northwest corner of Washington and Third Streets

1889–91
Abraham and Nellie Eddy Mathews House (destroyed)
Demetrius Frederick Charlton, architect
James Sinclair, masonry
Northeast corner of Bluff and Blaker Streets

Water Works (Marquette Maritime Museum)
Demetrius Frederick Charlton, architect
Furst, Jacobs and Company, contractor for superstructure
300 Lakeshore Boulevard
Portage Entry

1889–1900, 1907
Northern State Normal School, Longyear Hall of Pedagogy (South Wing) (destroyed 1990s)
Charlton, Gilbert and Demar, architects
1907 rebuilding, E. W. Arnold, architect
Kaye Avenue
Marquette

1890
Bice Pendill and Company
321 Division
Marquette variegated

1890–91
Marquette Opera House (destroyed)
Demetrius Frederick Charlton of Charlton and Gilbert, architect
Wilson and Moore, contractors
Washington Street between Front and Third Streets
Portage Entry

1890–92
John Munro and Mary Beecher Longyear House (Longyear Museum, Mary Baker Eddy Museum, Longyear at Fisher Hill) (1903–4 dismantled, transported, and reassembled in Brookline, Massachusetts)*
Demetrius Frederick Charlton, architect
Charles Van Iderstine, contractor
536 Arch Street

1891
Frank Carney Block
Andrew W. Lovejoy, architect
136 West Baraga Street

Fraternity Hall (Odd Fellows and Knights of Pythias Block)
A. W. Lovejoy and Hampson Gregory, architects
213 North Front Street
Rock River Brownstone Company
Rock River Brownstone Company white sandstone

1891–92
Lake Superior Hotel-Sanitarium (destroyed 1929)
Elijah E. Myers with Andrew W. Lovejoy, architects
J. B. Sweatt, contractor
In Mountain Park between Jackson Street and Blemhuber Avenue
Marquette variegated

Marquette County Savings Bank*
[Charles A.] Barber and [Earl W.] Barber, architects
Noble and Benson, contractors
107 South Front Street
Rock River Brownstone Company

Second Timothy Nester Block
Charlton and Gilbert, architects
Sinclair and Outerson, stonework contractors
121-129 West Washington Street

1892
Anton Zaaman Block
Hampson Gregory, architect/builder
229 West Washington Street

1893
Mrs. Peter White Phelps House
Charlton and Gilbert, architects
Charles Van Iderstine, contractor
433 East Ridge Street
Portage Entry

1893–94
Upper Peninsula Brewing Company (destroyed except for Charles Meeske House)
Andrew W. Lovejoy, architect
Meeske and Washington Streets and US 41 West
Marquette variegated

1894–95
Marquette City Hall (Old Marquette City Hall)*
Lovejoy and Demar, architects
Emil Bruce, contractor
220 Washington Street
Marquette

1901–2
Northern State Normal School, Peter White Hall of Science (North Wing) (destroyed 1975)
Charlton and Gilbert, architects
Kaye Avenue
Marquette

1902
Marquette Depot of the Duluth, South Shore and Atlantic Railroad
Demetrius Frederick Charlton [John D. Chubb?], architect
Lipsett and Gregg, contractors
Main Street between Front and Third Streets
Marquette

1902–4
Marquette County Courthouse*
Charlton and Gilbert, architects; Manning Brothers, landscape architects
Northern Construction Company, contractor
1984 restoration, Lincoln A. Poley, architect
400 South Third Street
Marquette and Portage Entry

1903–6
Baraga School (destroyed 1975)*
John D. Chubb, architect
Fourth and Rock Streets
Marquette brownstone salvaged from Grace Furnace

1913–15
Northern State Normal School, J. H. Kaye Hall (Central Building) (destroyed 1975)
Charlton and Kuenzli, architects
Herman Gundlach, contractor
Kaye Avenue

Ontonagon County

Ontonagon

1884–85
Ontonagon County Courthouse
Wolf, Jacobs and Company
Portage Entry

Minnesota

Saint Louis County

Duluth

1870
Clark & Hunter Blocks
Duluth Brownstone Company

1871?
Banning Block
Duluth Brownstone Company

1889
Masonic Hall and Opera House (Temple Opera Block)
McMillan and Stebbins, architects
203 East Superior Street
Arcadian Brownstone Company

Old City Hall and Jail (Architectural Resources, Inc.)
Oliver G. Traphagen, architects
126–132 East Superior Street
Portage Entry Quarries Company

1890
Fire Hall No. 1
Traphagen and Fitzpatrick, architects
First Avenue East and Third Street
Duluth Brownstone Company

1891
First Presbyterian Church
Traphagen and Fitzpatrick, architects
300 East Second Street
Bayfield or Fond du Lac

1891–92
Duluth Central High School*
Emmet S. Palmer and Lucien P. Hall, architects
Bounded by East Second Street, First Avenue East, East Third Street, and Lake Avenue
Iron River, Cranberry River, Flag River Brownstone Company, Arcadian Brownstone Company

1891–93
First Methodist Episcopal Church
(destroyed)
Weary and Kramer, architects
Third Avenue West and Third Street
Superior Brownstone Company, Bass Island
Brown Stone Company

1892
Oliver G. Traphagen House (Redstone, H. T.
Klatzky and Associates)
Traphagen and Fitzpatrick, architects
1511 East Superior Street
Portage Entry Quarries Company

1894–95
Duluth Board of Trade
Traphagen and Fitzpatrick, architects
301 West First Street
Portage Entry Quarries Company

1897
Duluth Station of the Chicago, Saint Paul,
Minneapolis and Omaha Railroad
Port Wing Quarry Company

c. 1900
C. H. Oppel Block
Oliver G. Traphagen, architect [?]
115 East Superior Street

1902
Chamber of Commerce
Reider and Warner
Portage Entry Quarries Company

George Crosby House
I. Vernon Hill, architect
2029 East Superior Street
Portage Entry

County Jail
Portage Entry Quarries Company

Duluth Public Library (Carnegie Library)
Adolph F. Rudolph, architect
101 West Second Street

Fowler Building
German and Brown, architects
Portage Entry Quarries Company

Lyceum Theater, Traphagen and Fitzpatrick
Portage Entry Quarries Company

Torrey Building
Portage Entry Quarries Company

1900–1910
Duluth Depot of the Duluth and Iron Range
Railroad (Endion Station, Convention and
Visitor's Center)
Tenbusch and Hill, architects
Lakewalk near Canal Park Entry Clock
Tower
Portage Entry Quarries Company

Jefferson School
Radcliff and Willoughby
Bounded by East Second Street, Tenth Avenue
East, East Third Street, and North Ninth
Avenue East
Portage Entry Quarries Company

Date unknown
Adam's Flats (post 1895)
Port Wing Quarry

Buffalo Flats
Cranberry River

City Pumping Station
Port Wing Quarry Company

Davis's Commission House
Port Wing Quarry Company

Elevator H
foundation
Port Wing Quarry Company

First National Bank
Duluth Brownstone Company

Wisconsin

Ashland County

Ashland

1886–88, 1901
Saint Agnes Roman Catholic Church (Our Lady of the Lake Catholic Church and School)
J. S. Chevigny, architect
Pernier and Chevigny, contractors
201 Lake Shore Drive (US 2)

1887
First National Bank (Ashland National Bank Block, Ellis-Main Building)
[Allan Darst] Conover and [Lew Forster] Porter, architects
101 Main Street West

Northern National Bank
West Second Street

Vaughn Library Building
Ashland Brownstone Company

1888
George W. Peck Block (NEMEC Insurance and Real Estate)
Conover and Porter, architects
311 Main Street East

1889
Security Savings Bank (Northern Wisconsin Abstract Company)
Conover and Porter, architects
Donald Brothers, stone cutters
212-214 Main Street West

Union Passenger Depot, Wisconsin Central Railroad (The Depot)*
Charles S. Frost, architect
400 Third Avenue West
Ashland or Prentice Brownstone Company

1889–92
Knight Block (destroyed, 1974)*
Conover and Padley, architects
Southeast corner of Main Street East and Ellis Avenue
Ashland Brown Stone Company

c. 1890
Bristol Block (Meyers Drugs)
315-317 Main Street West
Superior Brownstone Company

1891
C. E. Booth Sanitarium (The Platter Supper Club)
Barber and Barber, architects
Henry Asseltine, contractor
Prentice Park

1892
Ashland Opera Block (planned)
Cobb
Main Street West and Sixth Avenue West

First Presbyterian Church (destroyed)
Warren B. Hayes, architect
Corner of Third Street and Vaughn Avenue

1892–93
Wheeler Hall (Northern Wisconsin Academy), Northland College
1411 Ellis Avenue

1892–94
Ashland Post Office (Ashland City Hall)*
Willoughby J. Edbrooke, architect
601 Main Street West
Prentice Brownstone Company at Houghton Point

1894–95
L. C. Wilmarth Block Number One (All Pro Auto Parts)
L. C. Wilmarth, designer
Archie Donald and Company, masonry contractor
808 Main Street West
Ashland brownstone

1895
Emil Garnich Block

[Henry E.] Wildhagen and [Herman] Rettinghaus, architects
Archie Donald and Company, masonry contractor
Second Avenue
Excelsior Brownstone Company

Wilmarth School (First Assembly of God)
Henry E. Wildhagen, architect
901 Third Avenue West

1895–96
Masonic Temple Building
Charles McMillan, architect
Archie Donald and Company, masonry contractor
522 Main Street West

c. 1898
Wilmarth Block No. 2
200-210 Main Street West

1899
Beaser School (Cooperative Educational Service Agency)
Henry E. Wildhagen, architect
612 Beaser Avenue

1900
Ellis School (Ellis Park Business Center)
Henry E. Wildhagen, architect
310 Stuntz Avenue

c. 1900
Army Navy Store
501 Main Street West

1905 Ashland High School (destroyed)
Henry E. Wildhagen, architect
1000 Ellis Avenue

Bayfield County

Bayfield

1883–84
Bayfield County Courthouse (Apostle Islands National Lakeshore Headquarters)*
John Nader, architect
Cook and Hyde, contractor
415 Washington Avenue
Cook and Hyde, Basswood Island

1886
School (destroyed)
S. E. Jenison, architect
Pernier and Webster, contractor

1891, 1898
Holy Family Catholic Church
Brother Adrian Weaver, OFM, designer
232 North First Street

1892
Currie Bell Block (Roxanne's at Currie Bell)
2 North Second Street

1894–95
Lincoln High School (Bayfield Public Schools)
J. W. Hinckly, contractor
Bounded by Sweeny Avenue, Third Street, Lynde Avenue, and Fourth Street

c. 1900
Power House (Bayfield City Hall)
125 South First Street

1903–4
Carnegie Library (Bayfield Library)
Henry E. Wildhagen, architect
37 North Broad Street

1905
First National Bank (Antiques)*
Archie Donald, builder and stonemason
Northwest corner of North Second Street and Rittenhouse Avenue
Pike's quarry

Bayfield Township

Salmo

1894
Robinson D. Pike Summer House (Pinehurst Inn)

William Price, architect
Wisconsin 13, three miles southwest of Bayfield
Bayfield sandstone

1895–97
State Fish Hatcheries
H. P. Hadley? [Padley], architect
Wisconsin 13, on Chequamegon Bay two miles southwest of Bayfield at Pike's Creek
Pike's quarry

Iron River

1902–7
Iron River Bank (Jim's Meat Market)
South Main Street

Sand Island

1881
Sand Island Light

Port Wing

1903
School (destroyed 1980)
Henry E. Wildhagen, architect
John H. Foster and Sons, contractors
off Wisconsin 13
Port Wing quarry

Washburn

1888–89
Union Block-Meehan/Nelson (Chequamegon Books)
2 East Bayfield Street (Wisconsin 13)
Washburn Stone Company

1889
Bayfield County Bank (Counseling Center)
John Halloran, builder
14 East Bayfield Street (Wisconsin 13)

1889–90
Washburn State Bank (Washburn Historical Museum and Cultural Center)*
Conover and Padley, architects
1993 rehabilitation and addition
1 East Bayfield Street (Wisconsin 13)

1893–94
Walker School (destroyed 1947)*
T. Dudley Allen, architect
Washburn Stone Company

1894, c. 1960 and 1998 additions
Bayfield County Courthouse*
[Fremont D.] Orff and [Edgar E.] Joralemon, architects
John Halloran, contractor
117 East Fifth Street
Washburn Stone Company

1895
F. J. Meehan Clothing and Dry Goods Store
R. Orff, architect
Bayfield Street (Wisconsin 13)

c. 1900
Al Lien Block
Bayfield Street (Wisconsin 13)

1902
Saint Louis Catholic Church
Brother Adrian OFM, La Crosse
217 West Seventh Street

1903–5
Free Public Library (Washburn Public Library)*
Henry E. Wildhagen, architect
Sheridan and Swain, contractor
307 Washington Avenue
Washburn

Douglas County

Superior

1888, 1891
West Superior Hotel (Globe News)
1430 Tower Avenue (Wisconsin 35)

1890
Mayor Pattison House
foundation and basement
Arcadian Brownstone Company

Trade and Commerce Building (Superior City Hall)*
Clarence H. Johnston, architect
1409 Hammond Avenue
Southern Potsdam

1890–91
James Roosevelt Block (Roosevelt Terrace)
Carl Wirth, architect
Smith Brothers, contractors
1700-1714 Twenty-first Street

A. W. Stow Block
Northeast corner of Broadway and Hughitt Avenue

1890s
Commercial Block
1513-1515 Belknap Street (Business US 53)

1891
Church of the Redeemer (destroyed)
W. T. Towner, architect
Corner of West Fourth and Saint John Avenue
Arcadian Brownstone Company and Port Wing Quarry Company

T. W. Marble Double House (Brownstone Apartments)*
1311 John Avenue

John W. Palmer Building (planned)
Carl Wirth, architect
Corner of Tower Avenue (Wisconsin 35) and Seventh Street

Superior Depot of the Chicago, Saint Paul, Minneapolis and Omaha Railroad (destroyed)
Charles S. Frost, architect
C. W. Gindele, contractor
Portage Entry Quarries Company

Warehouse and Builders Supply (destroyed)
J. L. Rose, contractor
South side of Third Street between Tower (Wisconsin 35) and Ogden Avenues

1891–92
Breunig Block (destroyed)
Emil W. Ulrick, architect
Butler Brothers and M. P. Ryan, contractors
Tower Avenue (Wisconsin 35) and Broadway Street
Superior Brownstone Company

Minnesota Block (Board of Trade Building)*
Charles Coolidge Haight, architect
Southwest corner of Tower Avenue (Wisconsin 35) and Belknap Street (Business US 53)
Portage Entry trim

1892
Grand Republic Mill and Elevator
H. A. Hand, architect

William Listman Company Flouring Mill (Todd and Listman Mills)
Barnett and Record, contractors
Arcadian Brownstone Company

Saint Francis Xavier Catholic Church and Rectory
2316 East Fourth Street
Arcadian Brownstone Company

1892–93
Osburne-Burke-Chase Wholesale Grocery
R. J. Haxby, architect
Seymour and Hart, contractor
Hughitt slip

1893
Auger Block
Superior Brownstone Company

Northern Block
Conover, Porter and Padley, architects
Corner of Fifth Street and Thompson Avenue
Arcadian Brownstone Company

Waterman Block
Superior Brownstone Company

1901–2
Masonic Temple
Northeast corner of Belknap Street (Business US 53) and Hughitt Avenue

Superior Public Library (Carnegie Library)
Carl Wirth, architect
Southeast corner of Hammond Avenue and Twelfth Street

1909
Central High School (Central Junior High School)
1015 Belknap Street (Business US 53)

c. 1911
Amtrak Depot (Superior Antique and Art Depot)
933 Oakes Avenue

c. 1915
Bungalow
2115 Hughitt Avenue

Iron County

Hurley

1906
Saint Mary's Church
Northeast corner of South Fifth Avenue and Iron Street

APPENDIX 4

REPRESENTATIVE JACOBSVILLE AND BAYFIELD SANDSTONE BUILDINGS ELSEWHERE

Cities are listed in alphabetical order by state, country, or province.

City	Building	Dates	Architect	Quarry	Comments
Liverpool, England	McKenzie Bank Building			Portage Redstone	
Clinton, IA	Courthouse		J. Stanley Mansfield	Portage Entry and Marquette Raindrop/Portage Entry Quarries Co.	
Davenport, IA	First Presbyterian Church 316 Kirkwood Blvd.	1898–99	Gottshalk and Beadle, Frederick G. Clausen and Parke T. Burrows?	Variegated Marquette Raindrop/Portage Entry Quarries Co.	
Davenport, IA	United States Post Office				
Dubuque, IA	F. D. Stoutt House 1105 Locust St.	1890–91	W. W. Boyington and Co.	Variegated Portage Entry/Portage Entry Quarries Co.	
Chicago, IL	W. M. Crilley House Michigan Ave.		Flanders and Zimmerman	Marquette Raindrop/ Portage Entry Quarries Co.	
Chicago, IL	Cyrus McCormick House	1877	Cudell and Blumenthal		
Chicago, IL	Henry Furst House 505 Ashland Blvd.		Charles Furst (Furst and Randolph)	Variegated Marquette Raindrop/Portage Entry Quarries Co.	
Chicago, IL	Masonic Temple Forrest Ave. between 32nd and 33rd Streets			Portage Entry	
Chicago, IL	Potter Palmer House	1889	Angus and Gindele, builders	Bayfield/Prentice Brownstone Co.	
Chicago, IL	Tribune Building (destroyed 1902)	1872	Burling and Adler	Basswood Island/Strong, French and Co.	Dressed sandstone
Dubuque, IL	High School				
Monmouth, IL	Courthouse		O. W. Marble	Portage Entry/Portage Entry Quarries Co.	

City	Building	Dates	Architect	Quarry	Comments
Peoria, IL	City Hall		Reeves and Baillie	Portage Entry/Portage Entry Quarries Co.	
Peoria, IL	Webster School		Reeves and Baillie	Portage Entry/Portage Entry Quarries Co.	
Rockford, IL	United States Post Office			Portage Entry/Portage Entry Quarries Co.	
Rockford, IL	United States Post Office				
La Porte, IN	Courthouse		Mr. Tolin	Portage Entry/Portage Entry Quarries Co.	
New Orleans, LA	Canal Street Building			Portage Entry	
New Orleans, LA	Cotton Exchange			Portage Entry	
New Orleans, LA	First National Bank (Lottery Bank Building)			Portage Entry	
New Orleans, LA	Royal Building			Portage Entry	
New Orleans, LA	Sun Insurance Building			Portage Entry	
Bay City, MI	Masonic Temple	c. 1890		Rock River Brownstone Co.	
Charlotte, MI	Merchants' National Bank			Portage Entry/Portage Entry Quarries Co.	
East Lansing, MI	Morrill Hall, Michigan Agricultural College			Portage Entry	Red sandstone and red brick
Detroit, MI	Detroit Chamber of Commerce	1894–95	Spier and Rohns	Hermit Island/Excelsior Brownstone Co.	First two stories clad in sandstone
Detroit, MI	First Congregational Church Northeast corner of Woodward and Forest Aves.	1891	John Lyman Faxon	Portage Entry/Portage Entry Quarries Co.	
Detroit, MI	First Presbyterian Church Northeast corner of Woodward Ave. and Edmund Pl.	1889	Mason and Rice	Newport and Lake Superior Brownstone Co.	
Detroit, MI	Hammond Block	1890	Harry W. J. Edbrooke	Portage Entry	Solid rock-faced brownstone piers on first level

Detroit, MI	J. L. Hudson Co. Department Store Farmer St. and Gratiot Ave.	1891 (destroyed 1998)	Mortimer L. Smith	Stockton Island/Ashland Brown Stone Co.	
Detroit, MI	Masonic Temple	1894	Mason and Rice		
Detroit, MI	Parker Building			Portage Entry	
Detroit, MI	Schmidt Block	1894–95	Donaldson and Meier	Stockton Island/Ashland Brown Stone Co.	
Detroit, MI	A. L. Stephens House Southeast corner of Woodward and Ferry Aves.	1890	Mason and Rice	Portage Entry/Portage Entry Quarries Co.	
Detroit, MI	Telephone Building	1891		Ashland Brown Stone Co.	
Grand Rapids, MI	Peninsula Bank		Sidney J. Osgood		
Grand Rapids, MI	Trust Building (Michigan Trust Company Building) 40 Pearl St. NW	1891–92, 1913, 1920 additions, Henry Crowe	Solon S. Beman		Red brick above rusticated coursed red sandstone base
Grand Rapids, MI	Universalist Church		William G. Robinson		
Kalamazoo, MI	Kalamazoo Station of the Michigan Central Railroad (Kalamazoo Transit Center) 459 North Burdick St.	1887	Cyrus L. W. Eidlitz		Rusticated Portage Entry
Mount Clemens, MI	Bath House				
Muskegon, MI	Muskegon County Courthouse		Sidney J. Osgood	Portage Entry and Marquette Raindrop/Portage Entry Quarries Co.	
Muskegon, MI	Hackley Manual Training School			Patton and Fisher	
Muskegon, MI	Union Depot 586 West Western Ave.	1893–95	A. W. Rush Sidney J. Osgood	Marquette	Red brick and reddish brown variegated raindrop sandstone
Port Huron, MI	Harrington Hotel				
Saginaw, MI	Saginaw City Hall	(demolished)	Fred Hollister		

City	Building	Dates	Architect	Quarry	Comments
Traverse City, MI	Grand Traverse County Courthouse South Boardman	1898–99	Rush, Bowman, and Rush; after their dismissal C. M. Prall supervised construction	Portage Entry	Red Portage Entry sandstone base and trim on red brick
Minneapolis, MN	W. P. Arcking House 1800 Vine Pl.			Portage Entry Quarries Co.	
Minneapolis, MN	First Free (Free Will) Baptist Church	1891 (demolished c. 1930)	Long and Kees		
Minneapolis, MN	Globe Building Fourth and Cedar Sts.	1887–89 (demolished 1958–59)	Edward Townsend Mix		
Minneapolis, MN	Lumber Exchange Building			Prentice Brownstone Co.	
Minneapolis, MN	Minneapolis Public Library	1889 (demolished 1961)	Long and Kees		
Minneapolis, MN	Northwestern Guaranty Loan (Metropolitan) Building	1890 (demolished 1961)	Edward Townsend Mix	Portage Entry/Portage Entry Quarries Co.	Upper nine stories of fine rich brown sandstone
Saint Paul, MN	John H. Allen House 335 W. Summit Ave.	1892	J. Walter Stevens	Basswood Island/Superior Brownstone Co.	Randomly laid rusticated red sandstone
Saint Paul, MN	Blair Apartment House (Angus Hotel) 165 North Western Ave.	1887	Hermann Kretz and William H. Thomas	Bayfield brownstone	Red pressed brick trimmed with Bayfield sandstone
Saint Paul, MN	Peter F. Bowlin House 760 W. Summit Ave.	1892 (destroyed 1938)	Clarence H. Johnston	Basswood Island/Superior Brownstone Co.	
Saint Paul, MN	Germania Bank Building (St. Paul Building) 6 West Fifth St.	1888–90	J. Walter Stevens	Bayfield Brownstone Cc. or Cook and Hyde Portage Entry sandstone	Two primary facades of random rock-faced Bayfield stone resting on large solid sandstone blocks at the base. Secondary facades yellow brick
Saint Paul, MN	Germania Life Insurance (Guardian) Building Minnesota St. at Fourth St.	1888–89 (demolished 1970)	Edward P. Bassford		Granite and red sandstone
Saint Paul, MN	Manhattan Building			Portage Entry	

Saint Paul, MN	Merchants National Bank (McColl Building) 366-368 Jackson St.	1890	Edward P. Bassford		Rough and carved red sandstone with columns of polished gray granite
Saint Paul, MN	New York Life Insurance Company Building Minnesota St. at East Sixth St.	1887–89 (demolished 1967)	Babb, Cook, and Willard	Portage Entry/Portage Entry Quarries Co.	
Saint Paul, MN	Ryan Building			Portage Entry	
Saint Paul, MN	7th Street Building			Portage Entry	
Saint Paul, MN	Soo Building			Portage Entry	
Saint Paul, MN	Washington High School			Prentice Brownstone Co.	
Kansas City, MO	City Hall			Portage Entry	
Saint Louis, MO	Wainwright Building	1890–91	Louis Sullivan		
Fargo, ND	City Opera House				
Omaha, NE	City Hall			Portage Entry	
Omaha, NE	City Hall		Chas. F. Beindorf	Portage Entry/Portage Entry Quarries Co.	
Omaha, NE	Pacific Express Co. Building			Portage Entry/Portage Entry Quarries Co.	
Lincoln, NE	Parker Block			Superior Brownstone Co.	
Albany, NY	Hudson Canal Building				
Brooklyn, NY	Fire Department Headquarters			Portage Entry/Portage Entry Quarries Co.	
Buffalo, NY	Masonic Temple			Portage Entry	
New York, NY	Manhattan Savings Institution			Portage Entry/Portage Entry Quarries Co.	
New York, NY	7 Maiden Lane			Portage Entry	
New York, NY	U.S. Arsenal and Army Building			Portage Entry	
New York, NY	Waldorf-Astoria Hotel		J. H. Hardenberg	Portage Entry/Portage Entry Quarries Co.	
Cincinnati, OH	Carew Building				

City	Building	Dates	Architect	Quarry	Comments
Cincinnati, OH	Cincinnati City Hall	1888–93	Samuel Hannaford	Presque Isle/Ashland Brown Stone Co. or Houghton Point/Prentice Brownstone Co.	Missouri granite, pale yellowish brown Amherst, Ohio, sandstone and elaborately carved Bayfield brownstone
Cleveland, OH	Bradley Block			Portage Entry	
Cleveland, OH	Hollenden Hotel			Portage Entry	
Cleveland, OH	Lyceum Building			Portage Entry	
Cleveland, OH	Panorama Building			Portage Entry	
Cleveland, OH	Society for Savings Building		Burnham and Root	Portage Entry/Portage Entry Quarries Co.	Ten stories of Portage Entry above a two-story base of granite
Cleveland, OH	Stewart Chisholm House		C. F. Schweinfurth	Portage Entry/Portage Entry Quarries Co.	
Toronto, Ontario	Board of Trade		Janes and Janes	Portage Entry/Portage Entry Quarries Co	
Toronto, Ontario	Canadian Fire Insurance				
Toronto, Ontario	Royal Fire Insurance Building			Portage Entry	
Toronto, Ontario	Watson Building				
Pittsburgh, PA	Carnegie Office Building			Portage Entry/Portage Entry Quarries Co.	
Pittsburgh, PA	Dr. O. M. Edwards House		Sawers and Finglandt	Portage Entry/Portage Entry Quarries Co.	
Philadelphia, PA	Michigan Building at the Centennial Exposition of 1876	1876	Julius Hess	Marquette brownstone	Marquette brownstone foundation, Huron Bay roofing slate
Montreal, Quebec	New York Life Building (New York Mutual Life Building/New York)			Portage Entry	
Appleton, WI	High School		Herman (Henry?) Wildhagen	Portage Entry/Portage Entry Quarries Co.	

La Crosse, WI	Episcopal Church		M. S. Detweiler	Variegated Portage Entry/ Portage Entry Quarries Co.	
La Crosse, WI	John R. Paul House 1133 Cass St.	1893 (destroyed)		Basswood Island	Archie Donald, Ashland stone mason, cut and finished the stone
Madison, WI	Law Building, University of Wisconsin			Lake Superior	
Milwaukee, WI	T. A. Chapman Dry Goods Stone	1884–85 (destroyed)	Edward Townsend Mix	Washburn/Cook and Hyde	Bayfield sandstone and terra-cotta trimmed the pale yellow Milwaukee brick structure
Milwaukee, WI	Forest Home Cemetery Chapel		Ferry and Claus	Basswood Island	
Milwaukee, WI	Milwaukee County Courthouse	1870–73 (demolished 1976)	Leonard A. Schmidtner	Basswood Island/Strong, French and Co.	Cream city Milwaukee brick veneered with red sandstone
Milwaukee, WI	Plankinton Block	1885–86, 1888–89	Edward Townsend Mix	Basswood Island/Strong, French and Co.	Rock-faced sandstone
Milwaukee, WI	Plankinton House Hotel	1885–86	Edward Townsend Mix	Basswood Island , Cook and Hyde	Sandstone trimmed the cream city brick building
Milwaukee, WI	Saint Paul's Episcopal Church	1882–91	Edward Townsend Mix	Basswood Island/Cook and Hyde	Rock-faced deep red sandstone
Sheboygan, WI	High School			Superior Brownstone Co.	
Sparta, WI	Courthouse		M. E. Bell	Portage Entry/Portage Entry Quarries Co.	

Sources: Ernest Robinson Buckley, *On the Building and Ornamental Stones of Wisconsin*, Wisconsin Geological and Natural History Survey Bulletin No. 4, Economic Series no. 2 (Madison: State of Wisconsin, 1898); W. Hawkins Ferry, *The Buildings of Detroit* (Detroit: Wayne State University Press, 1980); *Houghton Daily Mining Gazette* Green Sheet, 18 October 1958; *Houghton Daily Mining Gazette*, 28 March 1983; *The Portage Entry Quarries Co.* A promotional brochure illustrating buildings in North America constructed of sandstone from the Portage Entry Quarries Company (Houghton, Mich.: Michigan Technological University Library Archives and Copper Country Historical Collections, n.d.); *The Prentice Brownstone Co.* (N.p, n.d.), prepared to fill large or small orders on short notice.

NOTES

Preface

1. Architectural tiles fired by Mary Chase Perry Stratton in the kilns of the Pewabic Pottery distinguish buildings throughout Michigan and the entire United States.
   Mary Chase Perry Stratton (1867–1961) founded the Pewabic Pottery Company in Detroit in 1904 and named it after the Pewabic copper mine near her hometown of Hancock. In collaboration with Horace James Caulkins, a Detroit manufacturer of dental products, Stratton systematically experimented with new firing techniques and chemical glazes in a revolutionary oil-burning kiln, which led her to the discovery of unique iridescent glazes.

Introduction

1. A. P. Swineford, *History of the Lake Superior Iron District Being a Review of its Mines and Furnaces for 1873* (Marquette, Mich.: Mining Journal Office, 1873), 79. Also in *Marquette Mining Journal,* 21 September 1872; and Swineford, *History and Review of the Copper, Iron, Silver, Slate and Other Material Interests of the South Shore of Lake Superior* (Marquette, Mich.: The Mining Journal, 1876), 274.
2. John W. Jochim, comp., *Michigan and Its Resources* (Lansing: Robert Smith and Co., State Printers and Binders, 1893), 81.
3. Increase A. Lapham's and James Hall's statements were quoted in "The Superior Brownstone Co.," *Ashland Daily Press,* Annual Edition, 1893, 64–65.
4. Milwaukee County Supervisors, Proceedings of the Board of Supervisors of Milwaukee County, July, September 1868, March 1869.
5. Alanson Sweet moved from Owasco, New York, to Milwaukee in 1835, and later to Kansas and Illinois. Sweet promoted many early plank roads and railroads, traded in grain and was active in politics (*Dictionary of Wisconsin Biography* [Madison: The State Historical Society of Wisconsin, 1960], 344). In Milwaukee he built homes, stores, ships, and the first steam elevator; on the lakes he built many lighthouses (A. T. Andreas, *History of Milwaukee, Wisconsin* [Chicago: Western Historical Co., 1881], 1176–77). Daniel Wells, Jr. (1808–1902) was a speculator, businessman, and congressman. Born in West Waterville, Maine, Wells speculated in Wisconsin land in 1835 and moved to Milwaukee the following year. Through investments in lumbering, transportation projects, and banks, he earned a fortune estimated at fifteen million dollars and a reputation as the wealthiest man in Wisconsin (Andreas, *History of Milwaukee, Wisconsin,* 369).
6. Frederic Heath, "The Milwaukee County Historical Society," *Wisconsin Magazine of History* (December 1947), 180; John Gregory, *Industrial Resources of Wisconsin*

(Milwaukee: Milwaukee News Company, Printers, 1872), 92; Jerome A. Watrous, *Memoirs of Milwaukee County* (Madison: Western Historical Association, 1909), 110; James S. Buck, *Pioneer History of Milwaukee, 1833–1841* (Milwaukee: Milwaukee News Company, 1876–86), 1:79, 2:52–57.

7. Alan Gowans, *Images of American Living* (New York: J. B. Lippincott Co., 1964), 350–52.
8. William H. Jordy, *American Buildings and Their Architects: Progressive and Academic Ideals at the Turn of the Twentieth Century* (Garden City, N.Y.: Doubleday and Co., 1972), 3:1–82, 378.
9. *The Brickbuilder* 4 (January 1895): 23.

Chapter 1

1. At Sault Sainte Marie, the Jacobsville formation lies beneath the Saint Marys River, and no outcrops were suited for quarrying. Yet here stand numerous buildings constructed of this light gray to reddish brown material. The stone was known locally as "canal rock" because it was excavated from the shipping locks canal between 1853 and 1855 and from the power canal of the Michigan Lake Superior Power Company in 1898. The most significant canal rock buildings are the Michigan Lake Superior Power Company Hydroelectric Plant (Edison Sault Electric Company, 1898–1902), Central Methodist Church, (1893–94), and Saint James Episcopal Church (1902–3).

   Douglass Houghton, Michigan state geologist, had observed this stone in specimens taken from the excavation for a mill race. He had warned against its use in the construction of the locks but had approved its use for ordinary walls (*Second Annual Report of the State Geologist,* Michigan House Doc. 8, 4 February 1839, 14–15). Nevertheless, the sandstone taken from the Saint Marys River was used for the protection walls of the canal and for stone fences. Carl Rominger observed that it decayed on exposure to frost and weather (C. Rominger, "Palaeozoic Rocks," *Geological Survey of Michigan. Upper Peninsula, 1869–1873* [New York: Julius Bien, 1873], 1:84).
2. Wm. Kenneth Hamblin, *The Cambrian Sandstones of Northern Michigan,* Michigan Department of Conservation, Geological Survey Division, Publication 51, 1958, 4–10, 15, 139, plate 1; David Marcel Hite, "Sedimentology of the Upper Keweenawan Sequence of Northern Wisconsin and Adjacent Michigan" (Ph.D. diss., University of Wisconsin, 1968), xix, figure 2, and 8–13; F. T. Thwaites, *Sandstones of the Wisconsin Coast of Lake Superior,* Wisconsin Geological and Natural History Survey, Bulletin 25, Scientific Series no. 8, 1912, 25–26 and 33–47.
3. John A. Dorr, Jr., and Donald F. Eschman, *Geology of Michigan* (Ann Arbor: University of Michigan Press, 1970), 91.
4. Hite, "Sedimentology," 193–94, 196; Richard W. Ojakangas and G. M. Morey, "Keweenawan Sedimentary Rocks of the Lake Superior Region: A Summary," *Geological Society of America Memoir* 156 (1982): 157–64.
5. J. C. Makens, J. P. Dobell, and A. D. Kennedy, "Study of Technical and Economic Aspects of an Expanded Stone Industry in Michigan," Michigan Department of Commerce, Michigan Office of Economic Expansion, Research Project no. 23 (Houghton: Institute of Mineral Research, Michigan Technological University, 1972), 11.
6. Hite, "Sedimentology," 137, 150, 155.
7. For earlier comparative data of Bass Island sandstone (1873–79), Portage Entry sandstone (1902), and L'Anse and Portage Entry sandstones (1895), see R. D. Irving and George A. Koenig in the J. W. Wyckoff Collection, Michigan Technological University Archives and Copper Country Historical Collections, Houghton, Mich.; and James B.

Knight (Commissioner of Mineral Statistics), Office of the Commissioner of Mineral Statistics, *Mines and Mineral Statistics* (Lansing: Robert Smith and Co., 1895).

8. Hite, "Sedimentology," 130, 134, 145–46, 148–50.
9. Hamblin, *Cambrian Sandstones of Northern Michigan,* 23–25, 42–43. Hite summarizes P. H. Oetking's explanation for these features: "Decomposition of organic matter with or without accompanying bacterial growth caused a low pH and reducing conditions. These conditions were most pronounced about the organic nucleus and decreased outward in a texturally homogeneous rock as a spherical front. The iron oxide would be removed in solution and precipitated beyond the influence of the organic core or pos sibly partially converted to a carbonate" ("Sedimentology," 129).
10. H. R. Schoolcraft, *On the Number, Value, and Position of the Copper Mines on the Southern Shore of Lake Superior,* 17th Cong., 2d sess., S. Doc. 5, 1822, 7–28.
11. Douglass Houghton, *Fourth Annual Report of the State Geologist,* Michigan House Doc. 27, 1841.
12. "Lake Superior in 1840," paper read before the Detroit Pioneer Society, January 1874, in Bela Hubbard, *Memorials of a Half-Century in Michigan and the Lake Region* (New York and London: G. P. Putnam's Sons, the Knickerbocker Press, 1888; reprint, Detroit: Gale Research Co., 1978), 25–50.
13. David D. Owen, *Report of a Geological Reconnaissance of the Chippewa Land District of Wisconsin,* 30th Cong., 1st sess., S. Exec. Doc. 57, v. 7, 1847, 54.
14. J. W. Foster and J. D. Whitney, *Report on the Geology and Topography of a Portion of the Lake Superior Land District, in the State of Michigan,* pt. 1, Copper Lands, U.S. 31st Cong., 1st sess., House Ex. Doc. 69, 1850, 110. Foster and Whitney commented also on the economic value of the granite of the Huron Islands in Marquette County. This "would afford beautiful and durable building material" and "is admirably adapted to the purposes of construction, separating readily into large tabular masses, and resisting the action of atmospheric agents" (ibid., pt. 2, The Iron Region, Together with the General Geology, U.S. 32d Cong., Spec. sess., Senate Ex. Doc. 4, 1851, 48–49, 82, 191).
15. "Lake Superior in 1840," 25–50.
16. Charles Lanman, *Adventures in the Wilds of the United States and British American Provinces* (Philadelphia: John M. Moore, 1856), 103–4. Of the eastern white sandstones at the Pictured Rocks, Lanman said: "The most conspicuous of them is perhaps three hundred feet high, but its most superb feature was demolished by a storm in the year 1816. That feature, according to a drawing in my possession, was an arch or doorway, fifteen feet broad and one hundred feet high, through which Indians were accustomed to pass with their canoes. In those days, too, from the crevices in these solid walls of whitish sandstone leaped forth beautiful cascades, and mingled their waters with those of the lake" (103).
17. *Ashland Press,* 21 and 25 December 1895.
18. *Marquette Mining Journal,* 12 April 1873.
19. H. G. Rothwell, "Methods of Stone Quarrying," Michigan Engineering Society, *Michigan Engineer's Annual for 1894,* 27–29; George P. Merrill, *The Collection of Building and Ornamental Stones in the U.S. National Museum: A Hand-Book and Catalog,* Annual Report of the Board of Regents of the Smithsonian Institute, pt. 2 (Washington, D.C.: GPO, 1886), 310–16, 319–21, 324–26; *Portage Lake Mining Gazette,* 18 July 1889; Oliver Bowles, *The Stone Industries* (New York and London: McGraw-Hill Book Co., 1939), 51–52, 83–92; John C. Smock, "Building Stone in New York," *Bulletin of the New York State Museum* 2 (September 1890): 302.
20. John H. Jacobs, "The History of the Life of John H. Jacobs Who Developed the Wonderful Sandstone Business of Lake Superior," TS., 1924, John. M. Longyear Papers, Marquette County Historical Society, Marquette, Mich.

21. Allan D. Conover, "Statistics and Information of Michigan and Wisconsin Quarries," in *Report on the Building Stones of the United States and Statistics of the Quarry Industry for 1880* (Washington, D.C.: GPO, 1884), 22–30.
22. W. H. Winchell, *Preliminary Report on the Building Stones, Clays, Limes, Cements, Roofing, Flagging and Paving Stones of Minnesota,* Geological and Natural History Survey of Minnesota, Miscellaneous Publications no. 8 (Saint Paul: Pioneer Press, 1880), 35.
23. By comparison, the crushing strength of granite ranges from 12,000 to over 47,000 pounds per square inch and indicated much greater strength. Its minimum strength exceeded the maximum strength recorded for Lake Superior sandstone. The crushing strengths in pounds per square inch of sandstone samples from seven sites in the Bayfield group were as follows: Presque Isle (Stockton Island), 2,001 to 6,244 pounds; Houghton, 4,166 to 6,502 pounds; Bass Island (Basswood Island), 4,310 to 5,126 pounds; Bayfield, 3,492 to 4,588 pounds; Port Wing, 1,658 to 5,498 pounds; and a site across the river from Fond du Lac, Minnesota, 4,668 to 5,931 pounds (Ernest Robertson Buckley, *On the Building and Ornamental Stones of Wisconsin,* Wisconsin Geological and Natural History Survey, Bulletin no. 4, Economic Series no. 2 [Madison: State of Wisconsin, 1898], 393–94). For results of crushing strength tests of Lake Superior sandstone conducted at the Michigan Mining School, see O. P. Hood to J. W. Wyckoff, 7 May 1902, Wyckoff Collection. The Hood test, which was limited to three specimens of red sandstone, presumably from Portage Entry, since Wyckoff was the superintendent of the Portage Entry Quarries Company, reported figures on crushing strength using approximately three-inch cubes (cf. Buckley's report on two-inch cubes). The three samples tested showed 9,804, 7,470, and 8,000 pounds per square inch, which is significantly greater than those reported in Buckley.
24. Office of the Commissioner of Mineral Statistics, *Mineral Statistics,* 1895, 144.
25. *Marquette Mining Journal,* 28 October 1871.
26. "The Sandstones of Lake Superior," *The Michigan Engineer's Annual Containing the Proceedings of the Michigan Engineering Society* for 1891, 51; *Portage Lake Mining Gazette,* 13 March 1895.
27. Kenneth Hudson has pointed out that two trends in the use of limestone were noticeable by the fourteenth century in Portland and Bath, England. For major projects architects shifted from choosing purely local materials to selecting the most suitable stone wherever it was found. As freestone grew in fashion for building, great demands to find quantities of stone ahead of quality had poor results. The same trends may be observed in the Lake Superior region in the nineteenth century. For the major projects architects selected the best of the red sandstone. As freestone grew in fashion after the Chicago fire of 1871, and as the demand exceeded the supply, inferior grades that lacked endurance and beauty were placed on the market. This sometimes gave the industry a bad reputation (Kenneth Hudson, *The Fashionable Stone* [Bath, Somerset, Great Britain: Adams and Dart, 1971], 14).
28. *Marquette Mining Journal,* 2 March 1878.
29. Ibid.
30. Office of the Commissioner of Mineral Statistics, *Mineral Statistics,* 194–95.
31. E. T. Malone to the Portage Entry Quarries Company, 28 May 1906 and E. T. Malone to J. W. Wyckoff, 12 December 1906, Wyckoff Collection. Roehm was a stone mason who did many works in Calumet, including his own house in Laurium.
32. J. (James) W. Wyckoff was born in Naples, New York. On 13 February 1865, Wyckoff enlisted in Company F, Ninth Infantry at Jackson and mustered out at Nashville, Tennessee, on 15 September 1865. On 3 January 1874, J. W. Wyckoff and Company advertised in the *Marquette Mining Journal* as contractors, builders, and dealers in building materials. In 1891, one year after his first wife, Emma, died at Jacobsville, he married Amelia C. Anderson of Marquette (*Marquette Daily Mining Journal,* 27 January 1890

and 24 February 1891). Wyckoff died in Houghton and was buried on 16 October 1934 in the family plot at Park Cemetery in Marquette.

33. Lowell A. Lamoreaux to W. C. [probably J. W.] Wyckoff, 5 January 1903; Frank L. Young and Carl E. Nystrom to Traverse Bay Redstone Company, 5 January 1903; W. P. Arcking to Wyckoff, 6 January 1903; James Young to Wyckoff, 7 January 1903; Jones and Hartley to Wyckoff, 10 January 1903; and H. K. Jennings to Wyckoff, 22 January 1903, Wyckoff Collection.
34. *The Portage Entry Quarries Co.,* Michigan Technological University Library Archives and Copper Country Collections (N.p., n.d., [c. 1900]). John H. Jacobs had employed Frank E. Parshley of Brooklyn to travel throughout the United States to photograph "some of the great structures" built of the stone from Lake Superior quarries. These photographs probably were used in the booklet (*Marquette Mining Journal,* 28 February 1891).
35. Peter Wolf, later attributed with introducing brownstone in Chicago, came to Chicago from Rehlingen, Rhinebreison, Germany, in 1847 (*Biographical Record: Biographical Sketches of Leading Citizens of Houghton, Baraga and Marquette Counties, Michigan* [Chicago: Biographical Publishing Co., 1903], 403–7).
36. Rominger, "Palaeozoic Rocks," 87–89, 90–91; *Marquette Mining Journal,* 21 November 1868.
37. *Marquette Mining Journal,* 28 May 1870. See also ibid., 22 October 1870, 9 September 1871.
38. Jacobs, "Life of John H. Jacobs."
39. Born in Huddersfield, Yorkshire, England, where his father owned one of the freestone quarries of the district, George Craig came to Montreal in 1854 with a group of masons to work on the great Victoria bridge across the Saint Lawrence River. Then he worked his way west, following the course of the Grand Trunk railroad as it was built, until, on reaching Detroit and hearing of the Lake Superior Country, he set out for Marquette. He arrived in Marquette in 1860. Here he worked on opening the Burt and Pendill quarries, opened a quarry at Laughing Whitefish, and managed the Hurley quarry at L'Anse (*Marquette Mining Journal,* 11 April 1891 and 29 October 1892).
40. Peter Wolf to S. P. Ely, Bill of Sale, 9 September 1872, Peter White Papers, Michigan Historical Collections, Bentley Library, University of Michigan, Ann Arbor. Jacobs claimed that the Marquette Board of Supervisors had assessed Wolf and Company's real property and personal property in the amount of $191,000, forcing the sale to the Marquette men, who organized the new company and acquired the property for $183,000 (Jacobs, "Life of John H. Jacobs").
41. Peter White (1830–1908) arrived in Marquette in 1849 and became an entrepreneur, politician, and businessman. Samuel P. Ely (1827–1900), born in Rochester, New York, and educated at Williams College, came in 1858 to Marquette, where he built a railroad from the iron mines to the lake and developed the Lake Superior, Republic, Washington, and Champion Mines. William Burt (1825–92) had assisted his father in the survey of the Lake Superior region in 1846. He came to Marquette in 1866 and speculated in real estate, iron mines, manufacturing, and brownstone and slate quarries. Frederick P. Wetmore (1813–83), born in Vermont, arrived in Marquette in 1864 and became a merchant. Sidney Adams (1831–1906), born in Herkimer, New York, came to Marquette in 1851 and invested in real estate and lumber. Henry R. Mather (1824–88), born in Middletown, Connecticut, came to Marquette in his twenties. He speculated in Marquette real estate, owned an interest in the Munising furnaces, and served as receiver in the land office and as county treasurer. (A. T. Andreas, *History of the Upper Peninsula of Michigan* [Chicago: Western Historical Co., 1883], 427–36; *National Cyclopaedia of American Biography,* 1896, s.v. "Pumpelly, Raphael," 3:362; *Dictionary of American Biography,* 1929, s.v. "Brooks, Thomas B," 3:89–91).

42. *Marquette Mining Journal,* 14 October 1876 and 1 December 1877.
43. For biographical information on members of the Burt family, see Silas Farmer, *History of Detroit and Wayne County and Early Michigan* (Detroit: Silas Farmer and Co., 1890), 1178–85; Andreas, *History of the Upper Peninsula,* 427–48; and *Memorial Record of the Northern Peninsula of Michigan* (Chicago: Lewis Publishing Co., 1895), 10–12.
44. A. T. Andreas, *History of Chicago: From the Early Period to the Present,* From the Fire of 1871 until 1885 (Chicago: A. T. Andreas Co., 1886), 3:84–85.
45. *Marquette Mining Journal,* 28 February 1891.
46. One account held that Wolf and Jacobs leased Lot One from George Craig and others who had purchased it from Ransom Shelden in 1871. Shelden had acquired it from the United States government in 1854 and had operated a trading post here (J. W. Wyckoff, "The Red Stone Industry of the Upper Peninsula," TS., Michigan Technological University Library Archives and Copper Country Historical Collections, Houghton, Mich.). Another account said that Henry Stafford, E. J. Mapes, and other Marquette men who had formed the Lake Superior Brown Stone Company in 1873 had purchased Lot One, a forty-three-acre wooded parcel extending one-half mile on the lakeshore east of the lighthouse reservation, and had leased for twenty-five years with the option to renew the lease for another twenty-five years, 294 acres adjoining it and including one-half mile of river frontage close to the entry where the water depth was sufficient for shipping. The lease contained easements giving the tenant the right to construct and install tramroads, docks, houses, and machinery, the right to cut and use timber, and the right to quarry and remove stone (*Marquette Mining Journal,* 22 February 1873; Lake Superior Brown Stone Company, Articles of Association, 25 February 1873, Michigan Department of Commerce, Corporation Division). From 1883 to 1898, the stockholders of the Lake Superior Brown Stone Company collected $150,232 in royalties for the stone extracted from Lot One (Table Showing Stone Quarried from Lot 1 Sec 19-53-32 from 1883 to 1898 inclusive, and Royalty Paid Thereon, Wyckoff Collection).
47. *Marquette Mining Journal,* 22 February 1873. For a description of the stone at Portage Entry, see also T. R. Brooks, "Iron Bearing Rocks (Economic)," *Geological Survey of Michigan. Upper Peninsula, 1869–1873,* pt. 1 (New York: Julius Bien, 1873), 58.
48. Portage Entry Quarries Company, Articles of Association, 28 February 1893, Illinois, Department of State; Table showing Stone Quarried from Lot 1, Sec 19-53-32 from 1883 to 1898 inclusive, and Royalty paid thereon and Table showing Stone Quarried from Sec. 18 from 1893 to 1898, inclusive in Wyckoff Collection; Wyckoff, "The Red Stone Industry of the Upper Peninsula."
49. Traverse Bay Red Sandstone Quarry, Rubble Stone Account, 1904–1905 and Sawed Stone Account, 1904–1905, Wyckoff Collection.
50. Office of the Commissioner of Mineral Statistics, *Mineral Statistics,* 1895, 144; 1896, 156; 1897, 158.
51. Andreas, *History of Chicago,* 3:48. Other investors included a man named Gindele, probably of Angus and Gindele, general contractors and builders. John Angus and Charles W. Gindele, descendants of Scotch and Bavarian stone workers, respectively, had established their business in 1881 and had built the Potter Palmer House (1889) in Chicago of Bayfield sandstone from the Prentice Brownstone Company (ibid., 83; Buckley, *On the Building and Ornamental Stones of Wisconsin,* 218).
52. *Marquette Mining Journal,* 11 March 1893.
53. Office of the Commissioner of Mineral Statistics, *Mineral Statistics,* 1896, 154–55; 1897, 158–59.
54. The quarries at Portage Entry employed Finns almost exclusively. At least twenty-seven of the thirty-five names on the time book of the Portage Entry Quarries Company for June 1908 were Finnish (Portage Entry Quarries Company Time Book, June 1908, Wyckoff

Collection). The Finns provided a great deal of the labor for the quarry industry.

55. *Marquette Mining Journal,* 17 January 1891. "The boring done shows conclusively that the sheet of stone which is now being quarried by Furst, Jacobs & Co. and the Portage Red Sandstone company underlies the entire Building Stone & Mineral Exploring company's tract, the greater portion of which, as is the case with the two properties mentioned, is a No. 2 quality. I succeeded, however, in locating a belt of No. 1 stone, beginning at a point 480 feet west of the northeast corner of said tract, this belt runs southeast by south, and cores taken out for a distance of about 900 feet at intervals of from 100 to 150 feet, the width of said belt being about 200 feet. It was my intention to continue testing the belt entirely across the property, as far as it extended, but as the drill had to go elsewhere I was not able to complete the work I had mapped out. I have no doubt, however, but that the belt of No. 1 stone extends nearly if not quite across the tract. . . . Comparing this No. 1 belt of stone with that which Furst, Jacobs & Co., are now working, and its value becomes more apparent, the total of which was 1,100 feet long, 225 feet wide with an average thickness of about 7 feet, giving 1,692,500 cubic feet of No. 1 stone, or only 252,000 feet in favor of the famous Furst, Jacobs & Co. quarry. Should the Building Stone & Mineral Exploring Co.'s belt prove to be 1,400 feet long it would give a total of 1,680,000 feet of No. 1 stone. Comment is not necessary." Drill samples demonstrated that "the splendid sheet of No. 1 stone on which the Furst-Jacobs and Malone companies are now working" extended upon it and in greater density than existed on the latter's properties when quarrying first had commenced (Number one stone sold for twenty-five cents per cubic foot). The Malone Brothers of Cleveland (and Chicago?) operated the Portage Entry Red Sandstone Company on a site (S18 T53N R32W) leased from Earl Edgerton of L'Anse.
56. *Pen and Sunlight Sketches of Duluth, Superior, and Ashland* (Chicago: Phoenix Publishing Co., 1892), 86.
57. Joseph Thomlinson to J. W. Wyckoff, 30 April 1908, Wyckoff Collection.
58. E. G. Malone to J. W. Wyckoff, 12 January 1909, Wyckoff Collection.
59. Thwaites, *Sandstones of the Wisconsin Coast of Lake Superior,* 45–46.
60. Quoted in *Bayfield Press,* 20 September 1884.
61. The newspaper listed production for 1892 as follows: Prentice Brownstone Company, 750,000 cubic feet; Superior Brownstone Company, 310,000 cubic feet; Ashland Brown Stone Company, 683,000 cubic feet; Smith and Babcock (Washburn Stone Company), 185,000 cubic feet; and Excelsior Quarry, 150,000 cubic feet (*Ashland Daily Press,* Annual Edition [1893], 81). The Prentice Brownstone Company and the Ashland Brownstone Company together produced more cubic feet than any company operating at Portage Entry.
62. G. B. Morey, "Geology of the Keweenawan Sediment near Duluth, Minnesota" (Master's thesis, Minneapolis, University of Minnesota, 1960).
63. *Ashland Press,* 20 July 1889. Prentice had leased earlier, in 1883, the quarry at Houghton Point to Captain H. Doherty, operator of the steamer *City of Ashland.* Doherty had planned to open it immediately and "get stone out for the Ashland trade," but there is no evidence that he did (*Bayfield Press,* 21 July 1883).
64. *Ashland Press,* 24 May 1890.
65. *The Prentice Brownstone Co.,* a promotional booklet for the company, n.d., 16. Michigan Technological University Library Archives and Copper Country Historical Collections, Houghton, Mich.
66. The *Bayfield Press* of 2 July and 3 September 1881 stated that Louis Lederly, superintendent of the Light House Construction Department, had cleared ten acres and had opened a red sandstone quarry on Sand Island to furnish stone for the lighthouses on Passage Island and Sand Island in Lake Superior and Belle Isle in the Detroit. There is no clear archaeological evidence that a quarry was located on Sand Island.

67. Buckley, *On the Building and Ornamental Stones of Wisconsin,* 190–93; *Ashland Press,* 27 August 1887, 20 July 1889, 24 May, 6 September 1890; *Ashland Weekly News,* 25 July 1888; *Washburn Times,* 18 September 1902; Prentice Brownstone Company, Articles of Association, 12 October 1888, State Historical Society of Wisconsin Archives, Madison; *Pen and Sunlight Sketches of Ashland,* 166–67.
68. R. C. Allen, *Mineral Resources of Michigan with Statistical Tables of Production and Value of Mineral Products for 1913 and Prior Years,* Michigan Geological and Biological Survey, Publication 16, Geological Series 13 (Lansing: Wynkoop, Hallenbeck, Crawford Co., 1914), 88.
69. Office of the Commissioner of Mineral Statistics, *Mineral Statistics,* 1899, 301.
70. Ibid., 1896, 154.
71. W. Cooper to J. W. Wyckoff, 28 January 1909, Wyckoff Collection.

Chapter 2

1. *Marquette Mining Journal,* 1 July 1871.
2. For descriptions of the Bay Furnace and the Bay Furnace Company, see *Marquette Mining Journal,* 11 September, 11 December 1869, 16 April 1870, and 2 June 1877. The company organized in 1869 with offices in Marquette. John Outhwaite of Cleveland served as president and treasurer, James Pickands as secretary, and R. A. White as superintendent (ibid., 13 November 1869).
3. Ibid., 4 June 1870.
4. *Marquette Lake Superior Mining Journal,* 14 November 1868. The building up of towns and the congestion in cities brought disastrous fires that drove townspeople to practice fire prevention and fire protection. Carl Bridenbaugh has noted that a series of great fires in Boston in 1653, 1676, and 1679 precipitated the adoption of fire codes, regulation, and defense. The great fire of 1676 that destroyed 150 houses motivated the General Court of Massachusetts to order that all new houses be built of "stone or bricke, & covered with slate or tyle" (Carl Bridenbaugh, *Cities in the Wilderness: The First Century of Urban Life in America 1625–1742,* 2d ed. [New York: Alfred A. Knopf, 1955], 56–61). Fires in cities and villages of the Lake Superior region brought the enactment of codes that restricted building materials to stone or brick.
5. Col. Chas. D. Robinson, editor of the (Green Bay) *Advocate,* in *Marquette Mining Journal,* 17 September 1870.
6. A series of consolidations with other railroads led to the Marquette and Ontonagon Railroad Company. The Iron Mountain Railroad, incorporated in 1855 as the area's first railroad, merged into the Bay de Noquet and Marquette Railroad in 1858, later to become the Marquette and Ontonagon Railroad, which extended its lines to Michigamme. The Marquette and Ontonagon Railroad Company had overcome great difficulties in constructing and operating a railroad in this remote and isolated part of the country. The erection of these utilitarian stone structures at Marquette gave the company the construction and maintenance facilities it needed.
7. *Marquette Lake Superior Mining Journal,* 21 November 1868.
8. In 1873 the Marquette and Ontonagon Railroad merged with the Houghton and Ontonagon Railroad Company. In 1886 the Marquette, Houghton and Ontonagon Railroad merged into the Duluth, South Shore and Atlantic Railroad. A complete inventory of the Duluth, South Shore and Atlantic Railroad properties indicates stone passenger depots at Sault Sainte Marie and Marquette (Henry E. Riggs, *Report on the Valuation of the Duluth, South Shore & Atlantic Railway as of June 30, 1911* [Ann Arbor, Mich.: n.p., 1911]).

9. Henry R. Mather had ties to Cleveland; he married the daughter of Morgan Hewitt, who came to Marquette to invest in mineral lands and became the first president of the Cleveland Iron Mining Company. Mather was related to the Mathers of Cleveland, developers of iron and steel industries in America and of the Cleveland Iron Mining Company, later the Cleveland Cliffs Iron Company in Ishpeming.
10. Peter White, John M. Longyear, Josiah G. Reynolds, E. H. Towar, and Alfred Kidder served as directors. Incorporators included Mather; Morgan L. Hewitt; David Murray, a grocer; and S. P. Ely, president of the Marquette, Houghton and Ontonagon Railroad and secretary of at least five early mining companies (*Marquette since 1864* [Marquette: Guelff Printers, 1864]).
11. Henry Lord Gay designed Mather's house on Ridge Street in 1867 and surely remodeled it in 1871. From 1880 to 1883, Gay lived in Europe and placed second among 293 entries in a competition at Rome for the national monument of King Victor Emanuel. He served as secretary of the Western Association of Architects, organized the Builder's and Trader's Exchange, and published *The Building Budget* (Andreas, *History of Chicago, 3:68–69; Journal of the American Institute of Architects,* 9 [August 1921]: 282).
12. *Marquette Mining Journal,* 23 November 1872, 15 March 1873, 10 January 1874.
13. Ibid., 10 August 1872.
14. Ibid., 16 September 1871.
15. Ibid., 7 February 1885; *First National Bank Centennial* (Marquette, Mich.?: n.p., 1964).
16. *Marquette Mining Journal,* 6 and 13 June 1885.
17. Ibid., 6 June 1874.
18. Struck designed the High School (1874–75), the Opera House (1874–77), the Ripka house (1875), the Traverse house (1875–77), and the steeples of the First United Methodist Church (1870–73). He supervised the construction of the Superior Building (1871–73) and Saint Paul's Episcopal Church (1874–76). By the building season of 1875, he was so busy he sent to Chicago for a draftsman to assist him (ibid., 24 April 1875).
19. Slate from the Huron Bay slate district, northwest of the Huron Mountains near Huron Bay and fifteen miles north of L'Anse, furnished a compatible roofing material for sandstone buildings. The slate was of fine texture and ranged in color from light steel and brown, to deep black, to bright red. Marquette investors in mineral lands and companies, aided by eastern and midwestern investors on whom they regularly relied to furnish needed development capital, formed companies to quarry, finish, and sell slate beginning in 1872, concurrently with the first activities of the sandstone industry. The slate companies included the Huron Bay Slate and Iron Company, the Stafford Slate Company, the Clinton Slate and Iron Company, Hurley's Huron Mountain Slate and Mining Company, and the Superior Slate and Mining Company. In 1881 several consolidated as the Michigan Slate Company.
20. *Marquette Mining Journal,* 10 July and 20 November 1875. Twenty-five years later, when the school burned to the ground, the people of Marquette still referred to it as "Marquette's handsome brownstone high school building" (ibid., 24 February 1900). The only changes immediately contemplated in the new school were to separate the boiler from the main structure, to place the entrances at the west and south to avoid a draft, to increase the size, and to add a manual training department.
21. *Marquette Mining Journal,* 17 April 1875.
22. Edward Townsend Mix, the son and grandson of sea captains, studied architecture in the office of Major Stone, a student of Ithiel Town, who was one of the early professional architects in New Haven. Mix came to Chicago in 1855, where he worked with W. W. Boyington for two years before moving to Milwaukee and establishing his own firm, which specialized in public buildings (Henry F. Withey, A.I.A., and Elsie Rathburn Withey, *Biographical Dictionary of American Architects (Deceased)* [Los Angeles: New Age

Publishing Co., 1956], 423–24, 577; Andreas, *History of Milwaukee,* 1499–1500; *The National Cyclopaedia of American Biography,* 1921, s.v. "Mix, Edward Townsend").

23. Mary Ellen Wietczykowski, Historic American Building Survey/Historic American Engineering Record, National Park Service, U.S. Department of the Interior, February 1969; Agnes Boyd Troeger, "History of Saint Paul's," [Milwaukee]: n.p., c. 1985.
24. Edward A. Schilling (1875–1952) was born in Auburn, New York, but spent his formative years in Ishpeming, Michigan. See Schilling File, American Institute of Architects (AIA) Michigan, Detroit.
25. *Marquette Mining Journal,* 12 April 1884.
26. Ibid., 29 May 1880.
27. Amos R. Harlow was born in Shrewsbury, Massachusetts. He platted the village of Marquette in 1850 and recorded it with revisions made by the Cleveland Iron Company on 8 September 1854. Harlow served as postmaster, township supervisor, highway commissioner, and justice of the peace. After the Marquette and Cleveland Iron Companies consolidated in 1853, Harlow diversified his business interests by investing in lumbering, farming, and real estate (*Memorial Record,* 5–10).
28. A. H. Holland and E. H. Dwight, *1886–7: Hand-Book and Guide to Marquette, L. S. Mich.* (Marquette: Mining Journal Book and Job Print, 1886), 12.
29. *Marquette Mining Journal,* 28 April 1888.
30. After inspecting sites at four locations, the Board of Commissioners selected a seventy-two-acre site in Marquette that the Marquette Business Men's Association then purchased and donated to the state. Later, additional land was acquired to gain full control of the prison's water supply. See E. P. Royce, P. A. Van Bergen, John Duncan, E. L. Mason, J. M. Wilkinson, and C. H. Hall of the Board of Commission of the State House of Correction and Prison, U.P., to the Honorable R. A. Alger, Governor of Michigan, 15 December 1886, Marquette County Historical Society, Marquette, Mich.
31. John H. Jacobs to J. M. Wilkinson, 13 September 1885, Marquette County Historical Society, Marquette, Mich.
32. Ibid.
33. Charlton went by the name of D. Fred rather than Demetrius Frederick Charlton (Michigan Department of Licensing and Regulation, D. Fred Charlton, Application to Practice Architecture, Filed at Michigan State Board for Registration of Architects, 13 October 1915).
34. C. Frederick Charlton (Chicago) to Kathryn Eckert, 7 November 1977; *Marquette Mining Journal,* 27 February 1892, 31 August 1918, and 27 and 28 January 1941.
35. *Marquette Mining Journal,* 25 January 1890.
36. James Munro Longyear, *Landlooker in the Upper Peninsula of Michigan* (Marquette, Mich.: Marquette County Historical Society, 1960).
37. Working from a photograph and a ground plan, in 1886–87 Gardner (1836–1915) had planned the addition and alterations to Longyear's frame house at 424 Cedar Street, originally built for A. P. Swineford by J. B. Sweatt in 1883 (*Marquette Mining Journal,* 17 July 1886 and 15 January 1887). Visits and contacts in Chicago probably familiarized Longyear with the residential designs of Samuel Atwater Treat (1839–1910) and with Frederick Foltz (1843–1916) for clients such as George Armour, Martin Ryerson, C. B. Farrell, and Charles Libby. Scott and Charlton had designed a bowling alley for the house on Cedar Street in 1889 (*Marquette Mining Journal,* 19 October 1889).
38. James Munro Longyear to Eugene C. Gardner, 14 February 1890, Longyear Papers, Marquette County Historical Society.
39. The development of the house can be traced in Longyear's instructions to the architects and in his reactions to their proposals. See James Munro Longyear to Eugene C.

Gardner, 22 February and 19 March 1890; Longyear to Frederick Foltz and Samuel Atwater Treat, 22 February and 7 March 1890, Longyear Papers.

40. Longyear to Eugene C. Gardner, 19 March 1890, Longyear Papers.
41. Charles C. Van Iderstine (1849–1926) became the leading contractor and builder in Marquette, erecting ninety houses and public buildings. He was the son of Jeremiah P. Van Iderstine, who supervised the construction of bridges and buildings along the Marquette and Ontonagon Railroad (Alvah L. Sawyer, *A History of the Northern Peninsula of Michigan: Its Mining, Lumber and Agriculture Industries* [Chicago: Lewis Publishing Co., 1911], 2:1038; *Marquette Mining Journal,* 21 January 1926).
42. For initial correspondence see James Munro Longyear to Richard Codman, 27 September 1890, Longyear Papers.
43. James Munro Longyear to Frederick Law Olmsted, 14 July 1891, Longyear Papers. Olmsted's landscape plan is summarized in the *Marquette Mining Journal,* 23 April 1892.
44. *Memorial Record,* 270.
45. The *Marquette Mining Journal* for 24 January 1903 explained that Longyear's decision to move the house to Brookline was "prompted by his desire to make his home in the east and the fact that no purchaser for the house has appeared." But the story is more dramatic. Mrs. Longyear had watched her son drown in Lake Superior, and after that she had no taste for Marquette. She moved to Brookline to be nearer Mary Baker Eddy. The *Marquette Mining Journal* for 9 November 1901 predicted that "the loss of the house would be deeply felt by every resident of the city."
46. *Marquette Mining Journal,* 5 July 1890. Nathan M. Kaufman married Mary Breitung in 1888.
47. *Men of Progress Embracing Biographical Sketches of Representative Michigan Men with an Outline History of the State* (Detroit: Evening News Association, 1900), 468; *Memorial Record,* 113–15; *Marquette Mining Journal,* 27 November 1917.
48. *Marquette Mining Journal,* 2 August 1890.
49. In 1891 Charles A. Barber and Earl W. Barber had offices in Duluth, Minnesota. The two previous years, from 1889 to 1890, and in subsequent years, beginning in 1892, the firm operated out of offices in Superior, Wisconsin. By 1892, however, Charles Barber had moved to Winnipeg. During its short stay in Marquette, the firm seems only to have designed the savings bank.
50. *Marquette Mining Journal,* 13 February 1892.
51. The First National Bank Building, the structure erected by Peter White after the fire of 1868, had served as temporary quarters for the bank until its building on Front Street could be built (*Marquette Mining Journal,* 9 February 1895; Minutes of the Common Council of the City of Marquette, 25 April 1893, City of Marquette, Marquette, Mich).
52. Minutes, Common Council, Marquette, 25 May 1893.
53. Ibid., 6 April 1893.
54. *Kalamazoo City and County Directory* (Detroit: R. L. Polk and Co., 1885–86, 1886–87).
55. Born in Vermont in 1864, Edward Demar studied and practiced architecture in Toronto, then moved West. He worked as a draftsman in Winnipeg, Manitoba, and in 1884 moved to Brandon and Regina, where he probably also worked as a draftsman. Demar and Lovejoy associated in the practice of architecture in November 1891. Edward Demar was considered "one of the best known draughtsman in the upper peninsula." He worked as the draftsman in J. B. Sweatt's office in Marquette during the height of Sweatt's extraordinary building career in the 1880s, then with B. H. Pierce and Company in Hancock, and, later, with E. E. Grip and Company of Ishpeming (*Marquette Mining Journal,* 7 November 1891; Sawyer, *History of the Northern Peninsula of Michigan,* 3:1267–68).
56. Andrew W. Lovejoy and Edward Demar, architects, Marquette City Hall, 1892, Drawings, 10 pages, held by Peter O'Dovero, Marquette, Mich.

57. Minutes, Common Council, Marquette, 2, 7, 16, 18 October 1893.
58. Ibid., 6 and 16 January 1894.
59. *Marquette Mining Journal,* 16 January 1894.
60. Ibid., 10 February 1895.
61. Ibid., 10 November 1894.
62. Ibid., 19 May 1894.
63. Ibid., 4 February 1895.
64. The Marquette County Courthouse was the inspiration for Robert Travers's Iron Cliffs County Courthouse, the courtroom of which held the murder trial of Lieutenant Frederick Marion in the film *Anatomy of a Murder.* Of the Iron Cliffs County Courthouse Travers said: "Few structures in the Peninsula presented a more startling pile of stone and slate and marble and mortar." Travers, *Anatomy of a Murder* (New York: Dell Publishing Co., Inc., 1958), 203.
65. *Marquette Mining Journal,* 6 December 1902.

Chapter 3

1. The school was named Michigan Mining School in 1885, Michigan College of Mines in 1897, Michigan College of Mining Technology in 1927, and Michigan Technological University in 1964. Act 70 authorized the board to obtain a suitable location for the school; lease or erect buildings; procure books, furniture, apparatus, and implements; and appoint a principal and teachers—all without incurring indebtedness. It instructed the board to obtain, establish, and classify a complete collection of the minerals of the Upper Peninsula and to obtain statistical and scientific information on them and all important discoveries and improvements in developing them. The legislature appropriated twenty-five thousand dollars from the general fund to carry out the provisions of the act.
2. James Wright was general superintendent of the Calumet and Hecla Mining Company; Thomas Chadbourne had studied and practiced law with Hubbell; John Senter worked for the Lake Superior Mining Company, acted as a blasting powder agent for E. I. DuPont and Company, and speculated in mine stock; John Forster served as superintendent of several copper mines and as construction engineer on the Portage and Lake Superior Ship Canal; Charles Wright was state geologist; and Graham Pope served as agent for the Franklin Mining Company.
3. In addition to serving as state senator, Jay Abel Hubbell (1829–90) served as district attorney for the Upper Peninsula, prosecuting attorney of Houghton and Ontonagon Counties, member of the United States House of Representatives, and circuit court judge for Baraga, Keweenaw, and Houghton Counties. Born in Avon, Oakland County, Michigan, and educated at the University of Michigan, Hubbell practiced law in Ontonagon before opening an office in Houghton in 1873 (*Copper Country Evening News,* 15 October 1900; *Memorial Record,* 381–83).
4. In addition to directing the Michigan Mining School, Marshman Edward Wadsworth served as state geologist of Michigan from 1888 to 1893. Born in East Livermore, Maine, in 1847, Wadsworth had taught at Boston Dental College, Harvard, and Colby, and had conducted geographical and geological surveys of New Hampshire and Minnesota (*National Cyclopaedia of American Biography,* 8 [New York: James T. White and Company, 1906], s.v. "Wadsworth, Marshman"; Minutes, Board of Control, Michigan Mining School, 17 August 1887, Michigan Technological University, Houghton, Mich.).
5. *Marquette Mining Journal,* 22 January 1887.

6. *Fifty-Fourth Annual Report of the Superintendent of Public Instruction of the State of Michigan, with Accompanying Documents for the Year 1890,* "Report of the Director and Treasurer, Michigan Mining School" (Lansing: n.p., 1890), 287–88.
7. *Marquette Mining Journal,* 4 February 1888.
8. *Houghton Daily Mining Gazette,* 12 December 1889.
9. *Fifty-Fourth Annual Report of the Superintendent of Public Instruction of the State of Michigan,* 296.
10. Albert Nelson Marquis, ed., *The Book of Chicagoans* (Chicago: A. N. Marquis and Co., 1911), 519; Henry F. Withey, A.I.A., and Elsie Rathburn Withey, *Biographical Dictionary of American Architects (Deceased)* (Los Angeles: New Age Publishing Co., 1956), 450–51; William Reichert, "The Work of Ottenheimer, Stern, and Reichert, Architects," *The Western Architect* 20 (December 1914): 127–29. Ottenheimer advertised offices in both Chicago and Houghton in the *Houghton Daily Mining Gazette* of 1899.
11. Sawyer, *History of the Northern Peninsula of Michigan,* 2:1012–14.
12. Juho Kustaa Nikander to Mackenzie, 29 April 1897, Suomi College Papers, Finnish-American Historical Archives, Finnish-American Heritage Center, Suomi College, Hancock, Mich.
13. The membership of the committee changed annually from 1890 to 1896, when the first board of directors was appointed. The board consisted of members of the consistory who served ex-officio and included the Reverends J. K. Nikander, K. L. Tolonen, Johan Back, and Kaarlo Huotari. In addition, this board included seven laymen, all business leaders in the Hancock and Calumet area: O. J. Larson, Victor Burman, Alex Leinonen, Jooseppi Riipa, J. H. Jasberg, August Pelto, and Andrew Johnson. In 1898 it appointed Carl Hensa (Minutes of the Suomi College Board of Directors, Suomi College Papers).
14. *Copper Country Evening News,* 31 May 1899.
15. *Houghton Daily Mining Gazette,* 18 December 1899.
16. Ibid., 13 January 1900.
17. Ibid., 21 January 1900.
18. Minutes, Board of Directors, Suomi College, 6 and 8 July 1898, Suomi College Papers; *Houghton Daily Mining Gazette,* 21 February 1900.
19. J. H. Jasberg to John D. Rockefeller, 25 March 1900, Suomi College Papers.
20. Minutes of the Common Council of the Village of Hancock, 2 February 1898, City Hall, Hancock, Mich.
21. *Copper Country Evening News,* 25 January 1898, 2 and 11 February 1898.
22. Ibid., 25 February 1898; Minutes, Common Council, Hancock, 23 February 1898.
23. Sawyer, *History of the Northern Peninsula of Michigan,* 3:1267–68.
24. Ibid., 1012–14; *Men of Progress,* 392.
25. Minutes, Common Council, Hancock, 11 April, 15 October 1898.
26. The council paid Charlton, Gilbert and Demar $385 (Minutes, Common Council, Hancock, 1 June 1898; *Copper Country Evening News,* 20 April and 23 August 1898).
27. *Copper Country Evening News,* 26 September and 4 November 1898, and 5 January 1899; *Marquette Mining Journal,* 30 January 1899.
28. *Marquette Mining Journal,* 30 January 1899.
29. For Ordinance No. 37, see Minutes, Common Council, Hancock, 4 October 1899. The area under the ordinance was "on the north half of Blocks 3, 7, 11, 13, 14, 15, and 17 and in the south halves of Blocks 4 and 8 and on all that land lying between Ravine Street and the Mineral Range Railroad track bounded on the south by Quincy Street and on the north by a line 120 ft north of Quincy St. extending from said Ravine St. to said Railroad track."
30. *Copper Country Evening News,* 17 February 1899. The people of nearby Calumet, struggling to convince their council that a town hall should be devoted to governmental

rather than social activities, pointed to the Hancock Town Hall and Fire Hall as a model of a sensible public building. The *Copper Country Evening News* wrote, "It would look as if the town council of Red Jacket [Calumet] need have gone no further than Hancock to have obtained an idea of what a Town Hall should be. The one there contains on the ground floor a fire engine room with stalls at the back for the teams. On the east side of the building is a wide staircase which leads to a very fine hall on the upper floor where the firemen give socials, it being reserved for their use, other entertainments being given in the public halls. On the east side of this hall is a very convenient office with a large safe for the village clerk (Justice Finn) and last but certainly not least a council chamber, *which is a council chamber,* and fitted up as such, the seats for the presiding officer and clerk being on raised dais, while the chairs for the councilmen are arranged within a semi-circular railing and not like the so-called Town Hall at Red Jacket, where everything had given way to the idea of an Opera House and assembly rooms as the space reserved for actual town purposes in the old building of the Red Jacket Town Hall is less than it was before as the part formerly used for the fire engine is to be turned into a banqueting hall and kitchen while the addition is pure and simple an Opera house" (4 May 1899).

31. Minutes of the Common Council of the Village of Red Jacket, 22 May 1885, 19 February, 19 March 1886, and 17 June 1887, Village of Calumet, Calumet, Mich.
32. *Copper Country Evening News,* 4 March 1897. For the account of the organization of the fire department, see Minutes, Common Council, Red Jacket, 2 November 1887.
33. Donald M. Scott, with offices in the O'Neil Block, Laurium, advertised as a civil engineer and architect. Charles W. Maass (1871–1959) practiced in Laurium for eleven years, seven of them in partnership with his brother, Fred A. H. Maass, and in Houghton for six years. Born in Green Bay, Wisconsin, he worked as an architect in Menominee for seven years before coming to the Copper Country in about 1896 (Charles W. Maass, Application for Registration to Practice Architecture, 15 October 1915, Michigan Department of Licensing and Regulation, Lansing).
34. *Copper Country Evening News,* 16 August 1898.
35. Ibid., 28 November 1898. For accounts of improvements to the old town hall and opera house, see *Copper Country Evening News,* 10 October and 21 November 1896 and 15 September 1897. For discussions that led up to the special election, see *Copper Country Evening News,* 2, 16, and 18 November 1898.
36. *Copper Country Evening News,* 28 November 1898.
37. *Copper Country Evening News,* 7 January 1899.
38. Ibid.
39. Ibid., 8 June 1899. Under the earlier plan, the city hall would be a model municipal building and the new theater the largest and finest north of the Straits of Mackinac. "There is but one theatre in Milwaukee, the Davidson, that will be its equal, and but one in Detroit that will be ahead of the new Calumet theater" (ibid., 2 October 1899).
40. Ibid., 13 and 15 March 1899; Minutes, Common Council Red Jacket, 11 March 1899.
41. *Copper Country Evening News,* 2 October 1899.
42. Ibid., 21 March 1900.
43. Ibid.
44. Kevin Harrington, "Photo Data Book for Ste. Anne's Church," *Historic American Building Survey,* National Park Service, U.S. Department of the Interior, July 1975.
45. *Copper Country Evening News,* 12 June 1900.
46. *Marquette Mining Journal,* 12 October 1895. The Johnson Vivian, Jr., House is a large house that combines Portage Entry red sandstone and shingles. It was designed by Demetrius Frederick Charlton. Located several blocks from the Roehm house, it was built for the son of a Cornish-born mine agent and investor who also established the J.

Vivian Mercantile Company (*Memorial Record,* 308–9; *Men of Progress,* 173).

47. In his deed to the trustees of the library, Paine set forth two purposes: "In memory of my mother, Sarah Sargent Paine, I have caused to be erected at Painesdale, in the Township of Adams, County of Houghton, and State of Michigan on land belonging to the Champion Copper Company, a stone building, primarily intended for use as a library building." Quoted from the deed for the library in Jane C. Lucchesi, *History of the Sarah Sargent Paine Memorial Library* (South Range, Mich.: n.p., 1978), 5.
48. Withey and Withey, *Biographical Dictionary of American Architects,* 199; Architect Files, American Institute of Architects (AIA) Michigan, Detroit, Michigan.
49. *Copper Country Evening News,* 16 September 1898.

Chapter 4

1. Henry Mower Rice, a Vermont-born Minnesota pioneer, U.S. Senator, and Indian commissioner, furthered the interest of the Minnesota Territory by procuring the extension of the preemption right over unsurveyed lands in Minnesota, the establishment of post offices and land offices, the extension of territorial roads, and by promoting the development of railroads and town sites, including Saint Paul. He summered in the Chequamegon Bay area (*Dictionary of American Biography,* 1929, s.v. "Rice, Henry Mower"; *Ashland Press,* 20 January 1894).
2. *History of Northern Wisconsin* (Chicago: Western Historical Co., 1881), 81.
3. The original two-story frame courthouse was built in 1873 at the corner of Broad Street and Frank Avenue (*Bayfield County Press,* 17 February 1883, 12 July 1884). Bicksler (1834–85) came to Bayfield from Virginia in 1856 to work as a carpenter for the Bayfield Land Company. In 1873 he moved to Ashland, where he also built the Chequamegon Hotel and other structures before opening a furniture store. Bicksler married the daughter of Elisha Pike, a Bayfield pioneer and the father of Robinson D. Pike, who became one of Bayfield's foremost citizens (*History of Northern Wisconsin,* 70; *Bayfield County Press,* 7 February 1885).
4. *Bayfield County Press,* 24 February 1883.
5. A pioneer, businessman, and village official, Robinson Deerling Pike (1838–1905) led the settlement and development of Bayfield. Born in Meadville, Pennsylvania, the son of Elisha Pike, he arrived with his family at La Pointe on Madeline Island in 1855 to homestead and operate on Pike's Creek a small water mill built by the American Fur Company ten years before but later abandoned. In 1856 Pike, Henry M. Rice, and others founded Bayfield. In the 1870s Pike established a shingle and saw mill; in the 1880s a planing mill, dockage facilities, boating business, and a sandstone quarry company. He promoted logging camps, better roads and docks, electricity, and telephones. People regarded Pike as progressive, energetic, and forceful and identified him with all moves to build up the community and surrounding country (Robinson Deerling Pike file, State Historical Society of Wisconsin, Madison, Wis.; *History of Northern Wisconsin,* 83–85; *Bayfield County Press,* 18 August 1905).
6. The enthusiasm for growth and development is expressed in the *Bayfield County Press* for 29 December 1883, which advertised in large print on its front page, "A FEW REASONS WHY PARTIES IN SEARCH OF NEW LOCATIONS FOR THE ESTABLISHMENT OF BUSINESS ENTERPRISE SHOULD PITCH THEIR TENTS IN THE HARBOR CITY."

    1. Because it is located on the largest, safest and most accessible harbor on the entire chain of lakes.
    2. The natural advantages it offers for the erection of the large docks, elevators, etc. are superior to all other points at the head of this lake.

3. It is the lake terminus of that great trunk line, the C., St. P., M. & O. RR [Chicago, St. Paul, Minneapolis, and Omaha Railroad].
4. The fall of '84 will witness the completion of the Northern Pacific R.R. and will have connections with the Great Northwest tapped by that road.
5. It is backed by country rich beyond ordinary comprehension in timber and minerals and finely adapted for agricultural purposes.
6. Its proximity to the Apostle islands renders it THE summer resort of the Lakes.
7. It is the seat of the largest fish packing industry on the lake.
8. It has immense quarries of the very best fire proof building stone.

7. The Houston County Courthouse at Caledonia City, Minnesota, was built to plans drafted by C. G. Maybury and Son of Winona, Minnesota, in 1883. In its simplicity and use of stone, the Old Bayfield County Courthouse bears a striking resemblance to the Houston County Courthouse (Minnesota Historic Properties Inventory Form for Houston County Courthouse, Site File, Minnesota Historical Society, Saint Paul, Minn.).
8. *Bayfield County Press,* 31 March 1883.
9. Ibid., 7 April 1883.
10. Minutes of the Bayfield County Board of Supervisors, 15 February, 1 March, 19 April 1883, Bayfield County Courthouse, Washburn, Wis.
11. *Bayfield County Press,* 5 May 1883. In a letter to the editor of the *Bayfield County Press,* E. M. Coffield of Trenton, New Jersey, also favored the use of sandstone because of its fire resistant quality: "I hope your people will decide to erect a court house of brown stone, with fire proof vaults. I was in Chicago when that city burned, and I can testify that the Lake Superior brown stone was the *only* stone that was not melted down" (*Bayfield County Press,* 28 April 1883).
12. *Madison Democrat,* 29 June 1919; Architect Files, Wisconsin State Historic Preservation Office, State Historical Society of Wisconsin, Madison.
13. *Madison Democrat,* 5 June 1883 (quoted in *Bayfield County Press,* 9 June 1883).
14. Minutes, Board of Supervisors, Bayfield County, 21 May 1883.
15. *Bayfield County Press,* 30 June 1883. Pike's quarry was first opened and developed in the spring of 1883. It is highly probable that only rough-cut stone was available from Pike's quarry at that time. On 3 March 1883 the *Bayfield County Press* noted that Captain R. D. Pike owned "a valuable undeveloped brownstone quarry on the bay near the village" for which he sought a developer. Pike offered a bonus of five hundred dollars to a responsible company that would open and operate it the coming season. The quarry fronted on the bay for at least one mile, and the Omaha Railroad traversed the property, both reducing the cost of putting in a system to transport the stone. By 12 January 1884 the *Bayfield County Press* reported lively activity at Basswood Island quarry. In 1883 Cook and Hyde leased and operated the Basswood Island quarry.
16. *Bayfield County Press,* 8 December 1883.
17. *Lumberman,* Bayfield, Wisconsin, 16 February 1885 (quoted in *Bayfield County Press,* 28 February 1885).
18. *Bayfield County Press,* 3 March 1905. For information on the events leading to the founding of the First National Bank see ibid., 5, 19, and 26 February 1904, 11 March 1904, and 4 August 1905.
19. The demand for Archie Donald's skills in cutting, finishing, and laying stone had grown with the emerging popularity of Bayfield sandstone in the 1890s. While cutting stone for a handsome new residence for lumberman John R. Paul in La Crosse, Donald's finishing technique was described as follows: "The stone is sawed and then rubbed together, making a very fine finish—in fact the very finest stone finish that can be procured." Donald's

skill and his extra attention to the project were predicted to make the residence one of the finest brownstone houses ever built (*Ashland Press,* 22 July 1894).

20. For discussions of the fire and rebuilding in stone, see *Bayfield County Press,* 22 September 1888 and *Washburn Times,* 12 January 1889.
21. *Pen and Sunlight Sketches of Duluth, Superior, and Ashland* (Chicago: Phoenix Publishing Co., 1892).
22. For a detailed account of the circumstances surrounding Probert's arrest, trial, and sentencing, see *Washburn News,* 10, 17 June, 15, 22 July, 16 September, 27 October 1893, and 1 January 1897.
23. The *Washburn News* for 7 July 1893 noted that among three plans submitted for the Bayfield County Courthouse at Washburn is "one from T. D. Allen, the school house architect." Although no contemporary document identifies Allen as the architect, the previous reference suggests that Allen was, indeed, the architect.
24. *Ashland Press,* 16 September 1893; *Washburn News,* 6, 20 May and 30 September 1893.
25. *Washburn News,* 24 November, 15 December 1893.
26. Ibid., 6 October 1893. The *Washburn News* had anticipated the prominence of the school at the very beginning of construction.
27. *Ashland Press,* 16 July 1892; *Washburn News,* 23 July, 29 October, 5 November, 3 and 12 December 1892.
28. *Washburn News,* 4 May 1894. See also, ibid., 6 May, 30 September 1893, 9 March 1894; Minutes, Board of Supervisors, Bayfield County, 25 May 1894.
29. Files of the Northwest Architectural Archives, University of Minnesota Libraries, Minneapolis, Minnesota; David Gebhard and Tom Martinson, *A Guide to the Architecture of Minnesota* (Minneapolis: University of Minnesota Press, 1977), 64, 169, 228–29, 323, 349; Site Files of the Minnesota Historical Society, Saint Paul.
30. James Bertam to W. H. Irish, 12 February 1903 (quoted in *Washburn News,* 2 July 1903).
31. *Washburn News,* 2 July 1903.
32. Ibid., 13 April 1905. For discussions of the planning, building, and opening of the library, see also ibid., 30 April, 21 May, 25 June, 2 July 1903, and 28 January 1904.
33. *Ashland Press,* 16 June 1888.
34. Ibid., 19 November 1887.
35. Ibid., 20 October 1888; *Ashland Weekly News,* 28 August 1889.
36. *Ashland Press,* 19 November 1887.
37. Blueprints for the Union Passenger Station, Chicago and Northwestern Transportation Company. Six pages of plans and elevations for the depot are on file at the Soo Line Railroad Company, Soo Line Building, Minneapolis, Minnesota. Withey and Withey, *Biographical Dictionary of American Architects,* 224.
38. *Ashland Weekly News,* 5 January 1889; see also *Ashland Press,* 15 June 1889.
39. *Superior Daily Leader,* 1 January 1892.
40. *Pen and Sunlight Sketches of Duluth, Superior, and Ashland,* 169.
41. Willoughby J. Edbrooke (1843–96) served as supervising architect of the United States Treasury Department from 1891 to 1892, during President Cleveland's administration. Emigrating from England as a youth, he began the practice of architecture in Chicago in 1867. He associated with Franklin P. Burnham between 1887 and 1891, while preparing plans for the Georgia State Capitol at Atlanta (Withey and Withey, *Biographical Dictionary of American Architects,* 1956], 189).
42. Myron H. McCord (1840–1908), a Wisconsin newspaperman, lumberman, and politician, served as congressman for one term, March 1889 to March 1891. Born in Ceres, Pennsylvania, McCord moved to Wisconsin in 1854, settling in Shawano where he logged and lumbered and published a newspaper. A Republican, McCord served as state senator in 1873–74 and state assemblyman in 1881, and as registrar of the United States

Land Office at Wausau in 1884–85. In 1897 he was appointed governor of Arizona Territory (*Dictionary of Wisconsin Biography* [Madison: The State Historical Society of Wisconsin, 1960], 245–46).

43. *Ashland Press,* 7 and 21 December 1889, 11 and 18 January, and 15 February 1890.
44. This law and a brief summary of the process to acquire the building appeared in the *Ashland Press,* 23 January 1892.
45. Appropriations and expenditures for the Ashland Post Office are stated in the U.S. Supervising Architect's *Annual Report, 1893,* U.S. Department of Treasury (Washington, D.C.: GPO, 1893).
46. *Ashland Press,* 4, 11 October, 29 November, 13 December 1890, and 31 January 1891. The site includes Lots 9, 10, 11, 12, 13, 14, 15, and 16 in the eastern portion of block 72, Vaughn Division of the city of Ashland. In 1892 the Office of the Supervising Architect of the Department of Treasury had under construction forty-six government buildings, purchased sites for twenty-two more, completed eighteen and spent a total of $6,741,286.71 (Glen Brown, "Government Buildings Compared with Private Buildings," *The American Architect and Building News,* 44, no. 954 [7 April 1894]: 2–9).
47. *Ashland Weekly Press,* 15 August 1891.
48. *Ashland Press,* 26 October 1895. Federal government building practices specified that the stone used in government buildings be "hard and durable, close-grained, of good texture, and free from defects and discolorations" and that a sample be sent to Washington, D.C. for approval. They required that the quarry from which the stone was proposed to be obtained be "fully opened and developed, and capable of supplying all the stone required within the time named by the bidder," and that its name be mentioned (James E. Blackwell, "The United States Government Building Practice," *American Architect and Building News,* 22, no. 601 [2 July 1887]: 3–7).
49. Willoughby J. Edbrooke to the Postmaster, Ashland, Wisconsin, 29 June 1892, Record of the Legislative and Natural Resources Branch, National Archives and Records Service, General Services Administration, Washington, D.C.
50. To alleviate these problems, the American Institute of Architects advocated a bill that permitted competition in design of government buildings with a jury of experts deciding the outcome on the merits of the plan. The bill was signed into law in 1893. The American Institute of Architects compared the cost of building, time of completion, cost of architect's services, business methods, and character of design in private buildings with that of government buildings. It demonstrated that first-class fireproof government buildings cost 62 percent more, took 300 to 500 percent longer to complete, and cost 3 percent more to design than private work. It pointed out that government buildings frequently were designed by subordinates or duplicated from drawings prepared for another project and built under the supervision of a man out of touch with the design. Supervising architects Edbrooke and O'Rourke took issue with the report and opposed the bill, claiming that "Government buildings are fully equal to municipal and private work of like magnitude in artistic effect, time of erection and cost of construction" (Brown, "Government Buildings Compared with Private Buildings," 2–9).
51. *Ashland Daily Press,* Annual Edition, May 1893. It was noted that work was begun on 15 August 1892 and was scheduled for completion by 15 November 1893, although the building was not completed until six months later. In comparing the government building in Ashland with one of similar cost, design, and date in Marquette, Michigan, it is evident that people in both communities desired a building executed in native stone but felt a sense of overriding pleasure at having secured so important a building at all. In 1881 Congress appropriated one hundred thousand dollars for a United States Custom House, Post Office, Court House, and Government Building at Marquette. The project took nearly ten years to complete. The treasury department purchased the site on

Washington Street in 1883. M. E. Bell and Will Freret supervised for the treasury department the preparation and revision of drawings for a Richardsonian Romanesque masonry structure, clearly derived from a source common to the Ashland building. Construction contracts were awarded in 1888, and the building was completed and occupied in 1890. The people of Marquette had preferred a building of native stone but got one with as much stone as the budget could afford. The *Marquette Mining Journal* for 24 March 1888 discussed the plans and specifications, then having just arrived from Washington:

> The material used will be pressed and moulded brick with terra cotta and rock-faced stone trimmings, although the specifications state that "bidders must state what additional amount, if any, they will charge to substitute, instead of pressed and moulded brick and terra cotta, sandstone or limestone, cut in the same shapes as nearly as possible, as the drawings provide the moulded brick." This makes it possible that the building may be constructed of Marquette or Portage Entry sandstone, though in any event it will be the handsomest public building in northern Michigan.

52. The *Superior Daily Leader* for 25 September 1892 explained that "the building is named in conformity with the plan adopted by the land company for naming its large blocks. Others are called after Wisconsin, New York, New Jersey, Massachusetts, Washington and Maryland." One group of early investors—Rensselaer R. Nelson, D. A. J. Baker, Daniel A. Robertson, and Henry M. Rice—were from Saint Paul.
53. John Henry Hammond was born in New York City, educated as a civil engineer at the University of Virginia, and studied horticulture in Spain and Switzerland (*Commemorative Biographical Record of the Upper Lake Region* [Chicago: J. H. Beers and Co., 1905], 5).
54. Alfred Morton Githens, "Charles Coolidge Haight," *Architectural Record,* 41 (April 1917), 367–69, *Dictionary of American Biography,* s.v. "Haight, Charles Coolidge;" Withey and Withey, *Biographical Dictionary of American Architects,* 255.
55. *Superior Daily Leader,* 17 November 1891. See also, *Superior Evening Telegram,* 17 November 1891.
56. H. H. Main of Madison, representative for the German-American Insurance Company, noted, "It is the best insurance risk I ever saw—it couldn't burn if it tried." *Superior Daily Leader,* 25 September 1892.
57. Ibid.
58. Ibid., 13 September 1891.
59. "Forum of Events: Death," *Architectural Forum* 66 (February 1937): 70.
60. *Superior Evening Telegram,* 27 May 1892, 8 November 1890.
61. *Superior Evening Telegram,* 16 May 1890.
62. Ibid., 8 November 1890.
63. The *Superior Leader* for 26 February 1893 reported that the sandstone came from Iron River; Ernest Robertson Buckley noted in his document published soon after the completion of the high school that the stone came from the Flag River Brown Stone Company and the Cranberry River quarry, located east of Superior in the westernmost portion of the Bayfield group (*On the Building and Ornamental Stones of Wisconsin,* 217).

BIBLIOGRAPHY

Archival Records

Papers and Collections

Hampson Gregory File. Marquette County Historical Society. Marquette, Mich.

John M. Longyear Papers. Marquette County Historical Society. Marquette, Mich.

Emil Lorch Papers. Michigan Department of State. Michigan Historical Center. State Archives of Michigan, Lansing.

Robinson Deerling Pike File. State Historical Society of Wisconsin, Madison.

Suomi College Papers. Suomi College Finnish-American Historical Archives. Finnish-American Heritage Center. Suomi College, Hancock, Mich.

Peter White Papers. Bentley Historical Library. Michigan Historical Collections. University of Michigan, Ann Arbor.

J. W. Wyckoff Collection. Michigan Technological University Archives and Copper Country Historical Collections. Houghton, Mich. The J. W. Wyckoff Collection contains important primary documents on the quarry industry, including the correspondence, quarry maps and drawings, time books and payroll sheets, ledgers, contracts, vouchers and miscellaneous papers of the superintendent of the Portage Entry Quarries Company.

Architectural Records

Architect Files. American Institute of Architects (AIA) Michigan. Detroit.

Madison, Wis. State Historical Society. Wisconsin Historic Preservation Office. Architect Files.

Minneapolis, Minn. University of Minnesota Libraries. Northwest Architectural Archives. Architect Files.

Saint Paul, Minn. Minnesota Historical Society. Site Files.

Minutes

Bayfield (Wis.) County Supervisors. Minutes of Meetings of the Board of Supervisors. 15 February–22 June 1883; 25 May–27 August 1894.
Hancock (Mich.) Village Council. Minutes of Meetings of the Common Council. 5 January 1898–4 October 1899.
Houghton, Mich. Michigan Mining School. Minutes of Meetings of the Board of Control. 15 July 1885–28 October 1890.
Marquette (Mich.) City Council. Minutes of Meetings of the Common Council. 6 April 1893–11 February 1895.
Marquette (Mich.) County Supervisors. Minutes of Meetings of the Marquette County Court House Building Committee. 30 April 1902–8 February 1905.
Michigan. Department of Licensing and Regulation. Minutes of Meetings of the Board for the Registration of Architects, 1915. 16 March 1917. State Archives of Michigan, Michigan Historical Center, Lansing.
Milwaukee (Wis.) County Supervisors. Proceedings of the Board of Supervisors of Milwaukee County. July, September 1868, March 1869.
Red Jacket. Minutes of Meetings of the Common Council. 22 February 1885; 1898–1900.
Saint Paul's Episcopal Church. (Marquette, Mich.) Minutes of Meetings of the Vestry. 21 August 1856, 13 April 1857, 12 March 1874–28 January 1889.

Typescripts

Barber, Ralph E. "The Years that Were 1851–1961 at Saint Paul's Episcopal Church of Marquette, Michigan." TS. [Marquette?], Mich.: n.p., n.d. Saint Paul's Episcopal Church, Marquette, Mich.
Boyer, Ken. Marquette Radio. TS. "Historic Highlights": "Stone Quarries at Laughing Whitefish Point and Salmon Trout," 23 March 1958; "City Stone Quarry on Ohio Street," 3 March 1963; "More Quarries and Rock Crushers," 6 October 1963.
Jacobs, John H. "The History of the Life of John H. Jacobs Who Developed the Wonderful Sandstone Business of Lake Superior." TS., 1924. Marquette County Historical Society. John M. Longyear Collection. Marquette, Mich.
———. "Life of John H. Jacobs." TS., n.d. Marquette County Historical Society. Marquette, Mich.
Wyckoff, J. W. "The Red Stone Industry of the Upper Peninsula." TS., n.d. Michigan Technological University Library Archives and Copper Country Historical Collections. Houghton, Mich.

Drawings

Charlton, Demetrius Frederick, and Edward Demar. "Drawings of the Mrs. Peter White Phelps House, 1893." Ink on linen drawings of plans and elevations. Held by David Drury, Marquette, Mich.

Frost, Charles S. Chicago and Northwestern Railroad Passenger Depot at Ironwood, Mich. Drawings. Held by Chicago and Northwestern Railroad Company Archives, Chicago, Ill.

Gregory, Hampson. Harlow Block. Ink on linen drawings of plans and elevations, n.d. [c. 1887]. Held by Alden Clark, Marquette, Mich.

Lloyd, Gordon W. "Proposed Finish to the Tower." St. Paul's Episcopal Church, n.d. [c. 1878]. Ink on vellum drawings. 4 pages. Held by Saint Paul's Episcopal Church, Marquette, Mich.

Lloyd, [Gordon W.], and Pearce. "St. Paul's Episcopal Church, Aug. 1873." Ink and watercolor on cloth-backed paper drawings of basement, first floor plan, tower elevation, cross section, side elevation. Held by Saint Paul's Episcopal Church, Marquette, Mich.

Lovejoy, [Andrew W.], and [Edward O.] Demar. "Marquette City Hall, 1893." Ink on linen drawings of south, east, north, and west elevations; foundation, first floor, second floor, and third floor plans; and sections on line A B and on line C D. 10 pages. Held by Peter O'Dovero, Marquette, Mich.

Maass, Charles W. Laurium (Mich.) Village Hall. Specifications of Work to be Done and Material to be Used in Remodeling the Laurium Village Hall. Laurium, Mich.

Wildhagen, Henry E. Free Public Library. Front Elevation 1903. Pen and ink watercolor. Held by Washburn Public Library, Washburn, Wis.

Other

The Portage Entry Quarries Co. A promotional brochure illustrating buildings in North America constructed of sandstone from the Portage Entry Quarries Company. N.d. [c. 1900], n.p. Michigan Technological University Library Archives and Copper Country Historical Collections. Houghton, Mich.

Michigan. Department of Commerce. Corporation Division. Articles of Incorporation. State Archives of Michigan. Michigan Historical Center, Lansing.

———. State Board of Registration for Architects, Professional Engineers and Land Surveyors. Applications for Registration to Practice Architecture. State Archives of Michigan. Michigan Historical Center, Lansing.

Red Jacket (Calumet, Mich.) Village Council. Fire Alarm Records.

Sault Sainte Marie, Mich. Edison Sault Electric Company Archives. Presidential Letter.

Newspapers

Ashland Daily Press
Ashland Press
Ashland Weekly News
Bayfield Press
Copper Country Evening News
Duluth Daily Tribune
Houghton Daily Mining Gazette
The Inland Ocean (Superior)
Iron Ore (Ishpeming)
Ironwood News Record
Lake Superior Journal (Marquette)
Lake Superior Mining Journal (Marquette)
Lake Superior News (Marquette)
Lake Superior News and Journal (Marquette)
Madison Democrat
Marquette Mining Journal
Milwaukee Daily Sentinel
Munising News
Portage Lake Mining Gazette (Houghton)
Sault Sainte Marie Evening News
Superior Daily Leader
Superior Evening Telegram
Torch Lake Times (Lake Linden)
Washburn Times

Government Documents

Allen, R. C. *Mineral Resources of Michigan with Statistical Tables of Production and Value of Mineral Products for 1913 and Prior Years.* Michigan Geological and Biological Survey. Publication 16, Geological Series 13. Lansing: Wynkoop, Hallenbeck, Crawford Co., 1914.

Buckley, Ernest Robertson. *On the Building and Ornamental Stones of Wisconsin.* Wisconsin Geological and Natural History Survey. Bulletin no. 4, Economic Series no. 2. Madison: State of Wisconsin, 1898.

Chamberlin, T. C. "Buildings Material." *Geology of Wisconsin: Survey of 1873–1879.* Vol. 1, pt. 3, 663–77. Madison: n.p., n.d.

Davidson, Edward S., Gilbert H. Espenshade, Walter S. White, and James C. Wright. *Bedrock Geology of the Mohawk Quadrangle, Michigan.* Geological Survey. U. S. Department of the Interior, 1955.

Eckert, Kathryn Bishop. "The Sandstone Quarries of the Apostle Islands: A Historical Narrative." Apostle Islands National Lakeshore, Midwest Region, National Park Service, U.S. Department of the Interior, 1985.

Foster, J. W., and J. D. Whitney. *Report on the Geology and Topography of a Portion of the Lake Superior Land District, in the State of Michigan.* Pt. 1, Copper Lands. U.S. 31st Cong., 1st sess., House Ex. Doc. 69, 1850. Pt. 2, The Iron Region, Together with the General Geology. U.S. 32d Cong., Spec. sess., Senate Ex. Doc. 4, 1851.

Gair, Jaco E., and Robert E. Thaden. *Geology of the Marquette and Sands Quadrangles, Marquette County, Michigan.* Michigan Department of Conservation. Geological Survey Division. Washington, D.C.: GPO, 1968.

Hamblin, Wm. Kenneth. *The Cambrian Sandstones of Northern Michigan.* Michigan Department of Conservation. Geological Survey Division. Publication 51. Lansing, 1958.

Hougton, Douglass. *Fourth Annual Report of the State Geologist.* Michigan House Doc. 27, 1841.

———. *Second Annual Report of the State Geologist.* Michigan House Doc. 8, 4 February 1839.

Irving, R. D. "General Geology of the Lake Superior Region." *Geology of Wisconsin. Survey of 1873–1879.* Vol. 3. Madison: Commissioners of Public Printing, 1880.

Jochim, John W., comp. *Michigan and its Resources.* Lansing: Robert Smith and Co., State Printers and Binders, 1893.

Makens, J. C., J. P. Dobell, and A. D. Kennedy. "Study of the Technical and Economic Aspects of an Expanded Stone Industry in Michigan." Michigan Department of Commerce. Michigan Office of Economic Expansion. Research Project no. 23. Houghton, Mich.: Institute of Mineral Research, Michigan Technological University, 1972.

Merrill, George P. *The Collection of Building and Ornamental Stones in the U.S. National Museum: A Hand-Book and Catalog.* Annual Report of the Board of Regents of the Smithsonian Institute, pt. 2. Washington, D.C.: GPO, 1886.

Michigan. Bureau of Labor and Industrial Statistics. *Fourth Annual Report.* "Convict Labor." Lansing: n.p., 1887.

———. Commissioner of Immigration. *Michigan and its Resources.* Frederick Morley, comp. Lansing: W. S. George and Co., 1881.

———. Commissioner of Mineral Statistics. *Annual Report of the Commissioner of Mines and Mineral Statistics,* 1877–1910. Lansing.

———. Department of State. "Upper Peninsula Resources." By R. A. Parker. In *Michigan and its Resources: Sketches of the Growth of the State, its Industries, Agricultural Productions, Institutions, and Means of Transportation: Description of its Soil, Climate, Timber, Financial Condition, and the Situation of its Unoccupied Lands; and a Review of its General Characteristics as a Home.* John W. Jochim, comp. Lansing: Robert Smith and Co., 1893.

———. *Geological Survey of Michigan. Upper Peninsula, 1869–1873.* Vol. 1. Pt. 1: T. B. Brooks, "Iron Bearing Rocks (Economic)." Pt. 2: Raphael

Pumpelly, "Copper-Bearing Rocks." Pt. 3: C. Rominger, "Palaeozoic Rocks." New York: Julius Bien, 1873.

———. *Mineral Resources of Michigan: With Statistical Tables of Production and Value of Mineral Products for 1914 and Prior Years.* Michigan Geological and Biological Survey. Publication 19, Geological Series 16. Lansing: Wynkoop, Hallenbeck, Crawford Company, 1915.

———. *Report of the Board of Commissioners of the State House of Correction and Branch of the State Prison in the Upper Peninsula.* Lansing: Darius D. Thorp, State Printer and Binder, 1889.

———. *Sixth Annual Report of the Bureau of Labor and Industrial Statistics.* Lansing: n.p., 1889.

———. [State House of Correction]. Upper Peninsula Branch. *Biennial Report of the Inspectors and Officers of the State House of Correction and Branch of the State Prison in the Upper Peninsula for the Two Years Ending June 30, 1892, 1894, and 1896.* Lansing: Robert Smith and Co., State Printers and Binders, 1893, 1894, 1897.

———. Superintendent of Public Instruction. *Fifty-Fourth Annual Report of the Superintendent of Public Instruction of the State of Michigan, with Accompanying Documents for the Year 1890.* "Report of the Director and Treasurer, Michigan Mining School." Lansing: n.p., 1890, 266–96.

Milwaukee, City of. Department of City Development. *Built in Milwaukee: An Architectural View of the City.* N.d. [c. 1973].

Owen, D. D. *Report of a Geological Reconnaissance of the Chippewa Land District of Wisconsin.* 30th Cong., 1st sess., S. Exec. Doc. 57, v. 7, 1847, 54.

Schoolcraft, H. R. *On the Number, Value, and Position of the Copper Mines on the Southern Shore of Lake Superior.* 17th Cong., 2d sess., S. Doc. 5, 1822, 1–28.

Smock, John C. "Building Stone in New York." *Bulletin of the New York State Museum* 2 (September 1890): 191–396.

Thwaites, F. T. *Sandstones of the Wisconsin Coast of Lake Superior.* Wisconsin Geological and Natural History Survey. Bulletin 25, Scientific Series no. 8. Madison: n.p., 1912.

U.S. Department of Agriculture. Forest Service. *Historic Place Restoration Feasibility Study: Bay Furnace, Hiawatha National Forest, Near Munising, Michigan.* By William Kessler and Associates, Inc., 15 December 1978.

U.S. Department of the Interior. Census Office. *Tenth Census.* Vol. 10: *Report on the Building Stones of the United States and Statistics of the Quarry Industry for 1880.* By Henry Gannett, George W. Hawes, F. W. Sperr, Thomas C. Kelly, and others. "Statistics and Information of Michigan and Wisconsin Quarries." By Allan D. Conover. Washington, D.C.: GPO, 1884.

———. National Park Service. Historic American Building Survey/Historic American Engineering Record. *Sault Sainte Marie: A Project Report.* By

Terry Reynolds et al. Washington, D.C.: U.S. Department of the Interior, 1982.

U.S. Department of Treasury. *A History of Public Buildings.* Washington, D.C.: GPO, 1901.

———. Supervising Architect. *Annual Report, 1893.* Washington, D.C.: GPO, 1893.

U.S. Records of the Legislative and Natural Resources Branch, National Archives and Records Service, General Services Administration. Washington, D.C.

Winchell, W. H. *Preliminary Report on the Building Stones, Clays, Limes, Cements, Roofing, Flagging and Paving Stones of Minnesota.* Geological and Natural History Survey of Minnesota. Miscellaneous Publications no. 8. Saint Paul: Pioneer Press, 1880.

Books

Andreas, A. T. *History of Chicago: From the Early Period to the Present.* Vol. 2: From 1858 until the Fire of 1871. Vol. 3: From the Fire of 1871 until 1885. Chicago: A. T. Andreas Co., 1885 and 1886.

———. *History of Milwaukee, Wisconsin.* Chicago: Western Historical Co., 1881.

———. *History of the Upper Peninsula of Michigan.* Chicago: Western Historical Co., 1883.

Arkell, W. J. *Oxford Stone.* London: Faber and Faber, 1947. Reprint, East Ardsley, Wakefield, Yorkshire: S. R. Publishers, 1970.

Armour, Robert E. *Superior, Wisconsin: A Planned City.* Superior, Wis.: Telegram Commercial Printing Division, 1976.

Art Work of the Lake Superior Region of Michigan. Oshkosh, Wis.: Art Photogravure Co., 1898.

Ashland and Environs: Ashland, Bayfield, Washburn and La Pointe, Picturesque and Descriptive. Neenah, Wis.: Art Publishing Co., 1888.

Ashland-Bayfield Bicentennial Committee. *Chequamegon: A Pictorial History of the Lake Superior Region of Ashland and Bayfield Counties.* Park Falls, Wis.: F. A. Weber and Sons, 1976.

Ashland City Directory. *Ashland, Wisconsin and St. Paul, Minnesota.* Detroit: R. L. Polk and Co., 1893.

Ashland, Wis., Illustrated. Milwaukee: Art Gravure and Etching Co., 1891.

Avery Obituary Index of Architects and Artists. Boston: G. K. Hall and Co., 1963.

Bayfield Historical Society. *Help Restore and Preserve the Heart of Bayfield's Heritage.* Bayfield, Wis.: n.p., n.d. [c. 1978].

Bayfield, Lake Superior: Early History, Situation, Harbor, etc.: Ocean Commerce, Mineral and Agricultural Resources, Rail Roads, Stage Roads, etc., Lumber, Fisheries, etc., Climate of Lake Superior, Pre-Emption Lands, Invitations to Settlers. N.p.: n.p., May 1858.

Beard's Directory and History of Marquette County: with Sketches of the Early History of Lake Superior. Detroit: Hadger and Bryce, Steam Book and Job Printers, 1873.

Benton, Marjorie F. *Historical Study of Chequamegon Bay Area.* Ashland, Wis.: American Association of University Women, Chequamegon Branch, 1972.

Biographical Record: Biographical Sketches of Leading Citizens of Houghton, Baraga and Marquette Counties, Michigan. Chicago: Biographical Publishing Co., 1903.

Bogue, Margaret Beattie, and Virginia A. Palmer, *Around the Shores of Lake Superior: A Guide to Historic Sites.* Madison, Wis.: University of Wisconsin Sea Grant College Program, 1979.

Bowles, Oliver. *The Stone Industries.* New York and London: McGraw-Hill Book Co., 1939.

Bradish, Alvah. *Memoir of Douglass Houghton.* Detroit: Raynor and Taylor, 1889.

Bridenbaugh, Carl. *Cities in the Wilderness: The First Century of Urban Life in America 1625–1742.* 2d ed. New York: Alfred A. Knopf, 1955.

Brisson, Steven C. T. "D. Fred Charlton's Architectural Practice and Design in the Upper Peninsula of Michigan, 1887–1918." Master's thesis, State University of New York College at Oneonta, 1992.

Brownstone and Bargeboard: A Guide to Bayfield's Historic Architecture. Madison, Wis.: University of Wisconsin Sea Grant Institute, 1980.

Bryant, William Cullen. *Picturesque America.* Vol. 1. New York: D. Appleton and Company, c. 1872–74.

Buck, James S. *Pioneer History of Milwaukee.* 4 vols. Milwaukee: Milwaukee News Company, 1876–86.

Burnham, Guy M. *The Lake Superior Country in History and in Story.* Ashland, Wis.: Ashland Daily Press, 1930.

Buvala, Fr. Medard. *History of St. Agnes Community 1873–1973 Celebrating One Hundred Years of Faith.* Ashland, Wis.: n.p., 1973.

Calumet, Houghton, Hancock and Laurium Directory. Detroit: R. L. Polk Co., 1910.

Calumet Village Centennial Committee. *Village of Calumet Michigan 1875–1975 Souvenir Centennial Book.* [Calumet, Mich.?]: n.p., n.d. [c. 1975].

Camden, Richard N. *Ohio—An Architectural Portrait.* Chagrin Falls, Ohio: West Summit Press, 1973.

Carter, James L., and Ernest H. Rankin, *North to Lake Superior: The Journal of Charles W. Penny 1840.* Marquette, Mich.: John M. Longyear Research Library, 1970.

Central United Methodist Church, Sault Ste. Marie. Sault Sainte Marie, Mich.: Historical Observance Committee for 135th Anniversary, 1833–1968, n.d. [c. 1968].

Chapple, John C. *Sketch of Ashland County, Wisconsin.* Iron Mountain, Mich.: C. O. Stiles, n.d. [c. 1903].

Charlton, [Demetrius Frederick], and [Edward O.] Kuenzli, *Some of Our Recent Work in the Upper Peninsula.* Quarter Century Souvenir Booklet. [Marquette, Mich.?]: n.p., 1914.

———. *Souvenir of a Few of Our Upper Peninsula Buildings.* [Marquette, Mich.?]: n.p., n.d. [c. 1907].

Church of the Redeemer. [Eau Claire, Wis.?]: Eau Claire Diocese, 1955.

The Church of the Transfiguration 1889–1979. Ironwood, Mich.: Nintieth Anniversary Booklet Committee, n.d. [c. 1979].

Citizens of Marquette. *Marquette, Mich. and Surroundings Illustrated.* Chicago: Pettibone Press, n.d. [c. 1896].

Coles, William A., ed. *Architecture and Society: Selected Essays of Henry Van Brunt.* Cambridge, Mass.: Belknap Press of Harvard University, 1969.

Commemorative Biographical Record of the Upper Lake Region. Chicago: J. H. Beers and Co., 1905.

Condit, Carl W. *The Chicago School of Architecture: A History of Commercial and Public Building in the Chicago Area 1875–1925.* Chicago and London: University of Chicago Press, 1964.

Craig, Lois. *The Federal Presence.* Cambridge, Mass., and London, England: MIT Press, 1978.

Dictionary of Wisconsin Biography. Madison: The State Historical Society of Wisconsin, 1960.

Donnan, Donald D., ed. *Marquette in 1900.* Chicago: Levytype Co., 1900.

Dorr, John A., Jr., and Donald F., Donald F. *Geology of Michigan.* Ann Arbor, Mich.: University of Michigan Press, 1970.

Dupree, A. Hunter. *Science in the Federal Government.* Cambridge, Mass.: Belknap Press of Harvard University, 1975.

Eaton, Leonard K. *American Architecture Comes of Age: European Reaction to H. H. Richardson and Louis Sullivan.* Cambridge, Mass., and London, England: MIT Press, 1972.

Eckert, Kathryn Bishop. *Buildings of Michigan.* New York: Oxford University Press, 1993.

———. "The Sandstone Architecture of the Lake Superior Region." Ph.D. diss., Michigan State University, 1982.

Fanning's Illustrated Gazetteer of the United States. New York: Ensign, Bridgman and Fanning, 1854.

Farmer, Silas. *The History of Detroit and Michigan or the Metropolis Illustrated.* Detroit: Silas Farmer and Co., 1884.

———. *History of Detroit and Wayne County and Early Michigan.* Detroit: Silas Farmer and Co., 1890.

First National Bank Centennial: A Hundred-Year History of the First National Bank & Trust Company of Marquette, Michigan. [Marquette, Mich.]: n.p., 1964.

First United Presbyterian Church: Historical Sketches for the 125th Anniversary Celebration 1854–1979. Sault Sainte Marie, Mich.: n.p., 1979.

Floyd, Margaret Henderson. *Architecture After Richardson: Regionalism before Modernism—Longfellow, Alden, and Harlow in Boston and Pittsburgh.* Chicago and London: The University of Chicago Press, 1994.

Gates, William B., Jr. *Michigan Copper and Boston Dollars.* Cambridge, Mass.: Harvard University Press, 1951.

Gebhard, David, and Tom Martinson, *A Guide to the Architecture of Minnesota.* Minneapolis: University of Minnesota Press, 1977.

Gowans, Alan. *Images of American Living.* New York: J. B. Lippincott Co., 1964.

Gregory, John. *Industrial Resources of Wisconsin.* Milwaukee: See-Bote, 1870. Reprint, Milwaukee: Milwaukee News Company, Printers, 1872.

Hanners, Arnold. *Ashland's Past: A Pictorial History.* Ashland, Ky.: Arnold Hanners, 1976.

Harris, Walt. *The Chequamegon Country 1659–1976.* Fayetteville, Ark.: W. J. Harris, 1976.

Herringshaw, Thomas W. *The Biographical Review of Prominent Men and Women of the Day with Biographical Sketches.* Chicago: Lewis Publishing Co., 1889.

History of Northern Wisconsin. Chicago: Western Historical Co., 1881.

Hite, David Marcel. "Sedimentology of the Upper Keweenawan Sequence of Northern Wisconsin and Adjacent Michigan." Ph.D. diss., University of Wisconsin, 1968.

Holland, A. H. and E. H. Dwight. *1886–7: Hand-Book and Guide to Marquette, L. S. Mich.* Marquette: Mining Journal Book and Job Print, 1886.

Holland, A. H. *The Marquette City Directory. 1891. Together with a Mining Directory of Marquette County.* Marquette, Mich.: Mining Journal, 1891.

Holleman, Thomas J., and James P. Gallagher. *Smith, Hinchman & Grylls.* Detroit: Wayne State University Press, 1978.

Holmio, Armas K. E. *Michigan in Suomalaisten Historia: Michigan Suomalaisten Historia-Suera.* Hancock, Mich.: Book Concern, 1967.

Hough, Jack L. *Geology of the Great Lakes.* Urbana: University of Illinois Press, 1958.

Houghton, J., Jr., and T. W. Bristol. *Reports of Wm. A. Burt and Bela Hubbard, Esqs. on the Topography and Geology of the U.S. Surveys of the Mineral Region of the South Shore of Lake Superior, for 1845.* Detroit: Charles Willcox, 1846.

Houghton County Directory, 1903–1904. Detroit: R. L. Polk and Co., 1904.

Hubbard, Bela. *Memorials of a Half-Century in Michigan and the Lake Region.* New York and London: G. P. Putnam's Sons, the Knickerbocker Press, 1888. Reprint, Detroit: Gale Research Co., 1978.

Hudson, Kenneth. *The Fashionable Stone.* Bath, Somerset, Great Britain: Adams and Dart, 1971.

Jordy, William H. *American Buildings and Their Architects: Progressive and Academic Ideals at the Turn of the Twentieth Century.* Vol. 3. Garden City, N.Y.: Doubleday and Co., 1972.

Jordy, William, and Ralph Coe, eds. *American Architecture and Other Writings by Montgomery Schuyler.* 2 vols. Cambridge, Mass.: Belknap Press of Harvard University, 1961.

Kolehmainen, John I., and George Hil. *Haven in the Woods.* Madison, Wis.: State Historical Society of Wisconsin, 1951.

Lankevich, George J., comp. and ed. *Milwaukee: A Chronological & Documentary History 1673–1977.* Dobbs Ferry, N.Y.: Oceana Publications, Inc., 1977.

Lankton, Larry, and Charles K. Hyde. *Old Reliable: An Illustrated History of the Quincy Mining Company.* Hancock: The Quincy Mine Hoist Association, Inc., 1982.

Lanman, Charles. *Adventures in the Wilds of the United States and British American Provinces.* Philadelphia: John W. Moore, 1856.

Longyear, John Munro. *Landlooker in the Upper Peninsula of Michigan.* Marquette, Mich.: Marquette County Historical Society, 1960.

Lucchesi, Jane C. *History of the Sarah Sargent Paine Memorial Library.* South Range, Mich.: n.p., 1978.

Marquette and Environs. N.p.: R. Acton, n.d. [c. 1887].

Marquette, Mackinac Island and the "Soo." Marquette, Mich.: B. F. Childs; and New York: Albertype Co., 1889.

Marquette, Mich. and Surroundings, Illustrated. [Marquette,] Mich.: The citizens of Marquette [1896?]; Chicago, Ill.: Pettibone Press.

Marquette, Michigan. Illustrated: Showing its Public Buildings, Some of its Private Residences, and Views in its Vicinity. With a Description of its Advantages for Business, and its Desirability as a Place of Residence. Milwaukee: Cramer, Aikens and Cramer, 1891.

Marquette Since 1864. Marquette: Guelff Printers, 1864.

Marquette "The Queen City of Northern Michigan": Its History, Industry, Natural Attractions. Marquette, Mich.: Guelff Printing Co., n.d. [c. 1925].

Marquis, Albert Nelson, ed. *The Book of Chicagoans.* Chicago: A. N. Marquis and Co., 1911.

———. *The Book of Detroiters.* Chicago: A. N. Marquis and Co., 1914.

McKee, Harley J. *Introduction to Early American Masonry.* Washington, D.C.: National Trust for Historic Preservation, 1973.

Memorial Record of the Northern Peninsula of Michigan. Chicago: Lewis Publishing Co., 1895.

Memorial Society of Michigan. *In Memoriam. Founders and Makers of Michigan: A Memorial History of the State's Honored Men and Women.* Detroit and Indianapolis: S. J. Clarke Publishing Co., n.d.

Mence, Edna. *L'Anse United Methodist Church—100 Years.* L'Anse, Mich.: Baraga County Historical Society, 1973.

Men of Progress: Embracing Biographical Sketches of Representative Michigan Men with an Outline History of the State. Detroit: Evening News Association, 1900.

Menominee City Directory. Detroit: R. L. Polk and Co., 1897–98, 1907–8.

Merila, Edith, ed. *Washburn Memories.* Shell Lake, Wis.: White Birch Printing, Inc., n.d. [c. 1982].

Merk, Frederick. *Economic History of Wisconsin During the Civil War Decade.* Madison: State Historical Society of Wisconsin, 1916.

Michigan State Gazetteer and Business Directory. Detroit: R. L. Polk and Co., 1881–1910.

Millett, Larry. *Lost Twin Cities.* Saint Paul: Minnesota Historical Society Press, 1992.

Milwaukee: The Beautiful. Official Souvenir of the Biennial Meeting of the General Federation of Women's Clubs. Milwaukee: Geo. H. Yerowine and Geo. W. Peck, Jr., 1900.

Monette, Clarence J. *The History of Jacobsville and its Sandstone Quarries.* Lake Linden, Mich.: Welden H. Curtin, 1976.

———. *Joseph Bosch and the Bosch Brewing Company.* Lake Linden, Mich.: Welden H. Curtin, 1978.

Moran, Stewart T., and Mrs. Stanley Pratt, comp. *St. James Episcopal Church: Historical Sketch 1832–1955.* Sault Sainte Marie, Mich.: n.p., n.d. [c. 1955].

Morey, G. B. "Geology of the Keweenawan Sediment near Duluth, Minnesota." Master's thesis, University of Minnesota, 1960.

Myers, Wallace Darlin, II. "Sedimentology and Tectonic Significance of the Bayfield Group. Upper Keewanean? (Upper Keewanean) Wisconsin and Minnesota." Ph.D. diss., University of Wisconsin, 1971.

Nikander, J. K. "Suomi-Opiston Johtajan Kertomus Suomi-Synoodin firfollisfo foufjelle wuonna 1899." In *Suomi-Opiston Luetto Lukuwonna 1898–1899.* Hancock, Mich.: Banen-Sanamain firjapinosja, 1900.

Noetzel, Benjamin D. *A History of Grace Methodist Church* [Houghton, Mich.?]: n.p., 1934.

O'Brien, William P. "Milwaukee Architect: Henry C. Koch." Master's thesis, University of Wisconsin-Milwaukee, August 1989.

O'Gorman, James F. *H. H. Richardson and His Office.* Cambridge, Mass.: Department of Printing and Graphic Arts, Harvard College Library, 1974.

———. *Living Architecture: A Biography of H. H. Richardson.* New York: Simon & Schuster Editions, 1997.

Osburn, C. S., ed. *The "Soo." Scenes in and About Sault Ste. Marie, Michigan.* Milwaukee: King, Fowle and Katz, n.d. [c. 1900].

Pare, Richard, ed. *Court House: A Photographic Document.* New York: Horizon, 1978.

Pen and Sunlight Sketches of Duluth, Superior, and Ashland. Chicago: Phoenix Publishing Co., 1892.

Perrin, Richard W. E. *The Architecture of Wisconsin.* Madison, Wis.: The State Historical Society of Wisconsin, 1967.

A Pilgrim's Guide to the Cathedral Church of St. Paul, the Apostle. Fond du Lac, Wis.: n.p., n.d. [c. 1970].

Purcell, Donovan. *Cambridge Stone.* London: Faber and Faber, 1967.

Ralph, Julian. *Along the Bowstring, or South Shore of Lake Superior.* N.p.: General Passenger Department of Duluth, South Shore and Atlantic Railway, n.d. [1893].

———. *Lake Superior along the South Shore from the Pen of Julian Ralph.* New York: American Bank Note Co., 1890.

Rezek, Rev. Antoine Ivan. *History of the Diocese of Sault Ste. Marie and Marquette: Containing a Full and Accurate Account of the Development of the Catholic Church in Upper Michigan with Portraits of Bishops, Priests and Illustrations of Churches Old and New.* 2 vols. Chicago: M. A. Donahue and Co., 1906–7.

Rickard, T. A. *The Copper Mines of Lake Superior.* New York and London: Mining and Engineering Journal, 1905.

Riggs, Henry E. *Report on the Valuation of the Duluth, South Shore & Atlantic Railway as of June 30, 1911: Explanation of Methods Used in the Valuation, a Complete Analysis of Unit Prices, the Presentation of Supporting Data Consisting of Actual Costs and Unit Prices Used on other Valuations and Some Comparisons with the 1900 and 1905 State Valuations.* Ann Arbor, Mich.: n.p., 1911.

Ritchie, James S. *Wisconsin and its Resources: With Lake Superior, its Commerce and Navigation. Including a Trip up the Mississippi, and a Canoe Voyage on the St. Croix and Brule Rivers to Lake Superior.* Philadelphia: Charles Desiller, 1857.

Sawyer, Alvah L. *A History of the Northern Peninsula of Michigan and Its People: Its Mining, Lumber and Agriculture Industries.* 3 vols. Chicago: Lewis Publishing Co., 1911.

Scott, James Allen. *Duluth's Legacy.* Vol. 1, *Architecture.* Duluth, Minn.: City of Duluth, Department of Research and Planning, 1974.

Skrubb, George N. *Marquette: The Story of Iron Ore. Marquette Centennial 1849–1949.* Marquette, Mich.: Stenglein Printers, 1949.

Smith, Peter. *Houses of the Welsh Countryside: A Study in Historical Geography.* London: HMSO, 1975.

The Soo: Scenes in and about Sault Ste. Marie, Michigan. Sault Sainte Marie, Mich.: N. C. Morgan, 1899.

A Souvenir of Ishpeming and Negaunee, Mich. Iron Mountain, Mich.: C. O. Stiles, n.d.

Spence, Clark C. *British Investments and the American Mining Frontier 1860–1901.* Ithaca, N.Y.: Cornell University Press, 1958.

The State of Wisconsin; Embracing Brief Sketches of its History, Position, Resources and Industries, and a Catalogue of its Exhibits at the Centennial at Philadelphia, 1876. Madison, Wis.: Atwood and Culver, 1876.

Steckbauer, W. E., and Albert Quade. *A Souvenir in Photogravure of the Upper Peninsula of Michigan. Calumet, Red Jacket, Laurium, Houghton, Hancock, Lake Linden, etc.* Brooklyn, N.Y.: Albertype Co., 1900.

Suomi-Opiston Albumi 1896–1906. Hancock, Mich.: Suomalais-Luteerilaisen Kustannus Lukkeen Kirjapaino, 1906.

Swierenga, Robert. *Pioneers and Profits: Land Speculation on the Iowa Frontier.* Ames, Iowa: Iowa State University Press, 1968.

Swineford, A. P. *Annual Review of the Iron Mining and Other Industries of the Upper Peninsula for the Year Ending December 13, 1881.* Marquette, Mich.: Mining Journal, 1882.

———. *History and Review of the Copper, Iron, Silver, Slate and Other Material Interests of the South Shore of Lake Superior.* Marquette, Mich.: Mining Journal, 1876.

———. *History of the Lake Superior Iron District Being a Review of its Mines and Furnaces for 1873.* Marquette, Mich.: Mining Journal, 1873.

Thurner, Arthur W. *Calumet Copper and People.* Hancock, Mich.: Book Concern, 1974.

Thurston, Robert N. *The Materials of Engineering.* Pt. 1. *Non- Metallic Materials: Stone; Timber; Fuels; Lubricants; etc.* New York: John Wiley and Sons, 1883.

Troeger, Agnes Boyd. *History of Saint Paul's.* [Milwaukee]: n.p., c. 1985.

Tyler, B. E. *Souvenir of the Copper Country Upper Peninsula of Michigan.* Houghton, Mich.: B. E. Tyler, 1903.

Van Rensselaer, Mariana Griswold. *Henry Hobson Richardson and his Works.* Boston, 1888. Reprint, New York: Dover, 1969.

Walsh, Margaret. *The Manufacturing Frontier: Pioneer Industry in Antebellum Wisconsin 1830–1860.* Madison: State Historical Society of Wisconsin, 1972.

Wargelin, John. *A Highway to America.* Hancock, Mich.: Book Concern, 1967.

Watrous, Jerome A. *Memoirs of Milwaukee County.* Madison: Western Historical Association, 1909.

Williams, Ralph D. *The Honorable Peter White. A Biographical Sketch of the Lake Superior Iron Country.* Cleveland: Penton Publishing Co., 1905.

Withey, Henry F., A.I.A., and Elsie Rathburn Withey. *Biographical Dictionary of American Architects (Deceased).* Los Angeles: New Age Publishing Co., 1956.

Wood, Ike. *One Hundred Years at Hard Labor: A History of Marquette State Prison.* Au Train, Mich.: Avery Color Studios, 1985.

Works Progress Administration, comp. *Michigan: A Guide to the Wolverine State.* New York: Oxford University Press, 1941–49.
———. *Minnesota, a State Guide.* New York: Viking, 1938. Reprint, St. Paul: Minnesota Historical Society Press, 1985.
———. *Wisconsin: A Guide to the Badger State.* New York: Hastings House, 1941.
Zimmermann, H. Russell. *Magnificent Milwaukee: Architectural Treasures 1850–1920.* Milwaukee: Milwaukee Public Museum, 1987.

Articles

Allen, R. C. , and Helen Martin. "A Brief History of the Geological and Biological Survey of Michigan: 1837 to 1872." *Michigan History* 6, nos. 2–3 (1922): 675–750.
Blackwell, James E. "The United States Government Building Practice." *American Architect and Building News* 22, no. 601 (2 July 1887): 3–7.
Brown, Glen. "Government Buildings Compared with Private Buildings." *American Architect and Building News* 44, no. 954 (7 April 1894): 2–9.
Clarke, Robert E. "Notes from the Copper Region." *Harper's New Monthly Magazine* 6 (March/April 1853): 433–48, 577–88.
Cooper, James B. "Historical Sketch of Smelting and Refining Lake Copper." *Proceedings of the Lake Superior Mining Institute* 7 (5–9 March 1901): 43–49.
Eaton, Leonard K. "John Wellborn Root and the Julian M. Case House." *Prairie School Review* 9 (1972): 18–22.
Eckert, Kathryn Bishop. "Made from the Rock They Stand On." *Michigan History Magazine* 78, no. 6 (November/December 1994): 62–69.
"Forum of Events: Death." *Architectural Forum* 66 (February 1937): 70 (supplement).
Githens, Alfred Morton. "Charles Coolidge Haight." *Architectural Record* 41 (April 1917): 367–69.
Heath, Frederic, "The Milwaukee County Historical Society." *Wisconsin Magazine of History* 31, no. 2 (December 1947): 178–85.
Illinois Society of Architects Monthly Bulletin 18 (April–May 1934): 7.
Julien, Alexis A. "The Decay of the Building Stones of New York City." *Transactions of the New York Academy of Sciences* (29 January 1883): 67–78; (30 April 1883): 120–38.
"New Water Power Plant at Sault Ste. Marie, Mich." *Engineer* 39, no. 16 (15 August 1902): 549–51.
Ojakangas, Richard W., and G. B. Morey. "Keweenawan Sedimentary Rocks of the Lake Superior Region: A Summary." *Geological Society of America Memoir* 156 (1982): 157–64.

Peterich, Gerda. "Cobblestone Architecture of Upstate New York." *Journal of the Society of Architectural Historians* 15 (May 1956): 12–18.

Pettee, W. H. "Notes on Building Stones." *The Michigan Engineer's Annual. Containing the Proceedings of the Michigan Engineering Society for 1889,* 50–54.

Reichert, William. "The Work of Ottenheimer, Stern, and Reichert, Architects." *Western Architect* 20 (December 1914): 127–29.

Rothwell, H. G. "Methods of Stone Quarrying." Michigan Engineering Society. *The Michigan Engineer's Annual. Containing the Proceedings of the Michigan Engineering Society for 1894,* 27–29.

Rothwell, Henry G. "The Sandstones of Lake Superior." *The Michigan Engineer's Annual. Containing the Proceedings of the Michigan Engineering Society for 1891,* 48–57.

Russell, James. "Peter White." *Michigan History* 6, No. 4 (1922): 296–314.

"The 'Soo' Water Power." *Engineering Record* 48, no. 13 (23 July 1898): 161–62.

"The Water-Power Plant of the Michigan-Lake Superior Power Company at Sault Ste. Marie." *Engineering News* (25 September 1902): 226–27.

INDEX

Titles in the Great Lakes Books Series

Freshwater Fury: Yarns and Reminiscences of the Greatest Storm in Inland Navigation, by Frank Barcus, 1986 (reprint)

Call It North Country: The Story of Upper Michigan, by John Bartlow Martin, 1986 (reprint)

The Land of the Crooked Tree, by U. P. Hedrick, 1986 (reprint)

Michigan Place Names, by Walter Romig, 1986 (reprint)

Luke Karamazov, by Conrad Hilberry, 1987

The Late, Great Lakes: An Environmental History, by William Ashworth, 1987 (reprint)

Great Pages of Michigan History from the Detroit Free Press, 1987

Waiting for the Morning Train: An American Boyhood, by Bruce Catton, 1987 (reprint)

Michigan Voices: Our State's History in the Words of the People Who Lived It, compiled and edited by Joe Grimm, 1987

Danny and the Boys, Being Some Legends of Hungry Hollow, by Robert Traver, 1987 (reprint)

Hanging On, or How to Get through a Depression and Enjoy Life, by Edmund G. Love, 1987 (reprint)

The Situation in Flushing, by Edmund G. Love, 1987 (reprint)

A Small Bequest, by Edmund G. Love, 1987 (reprint)

The Saginaw Paul Bunyan, by James Stevens, 1987 (reprint)

The Ambassador Bridge: A Monument to Progress, by Philip P. Mason, 1988

Let the Drum Beat: A History of the Detroit Light Guard, by Stanley D. Solvick, 1988

An Afternoon in Waterloo Park, by Gerald Dumas, 1988 (reprint)

Contemporary Michigan Poetry: Poems from the Third Coast, edited by Michael Delp, Conrad Hilberry and Herbert Scott, 1988

Over the Graves of Horses, by Michael Delp, 1988

Wolf in Sheep's Clothing: The Search for a Child Killer, by Tommy McIntyre, 1988

Copper-Toed Boots, by Marguerite de Angeli, 1989 (reprint)

Detroit Images: Photographs of the Renaissance City, edited by John J. Bukowczyk and Douglas Aikenhead, with Peter Slavcheff, 1989

Hangdog Reef: Poems Sailing the Great Lakes, by Stephen Tudor, 1989

Detroit: City of Race and Class Violence, revised edition, by B. J. Widick, 1989

Deep Woods Frontier: A History of Logging in Northern Michigan, by Theodore J. Karamanski, 1989

Orvie, The Dictator of Dearborn, by David L. Good, 1989

Seasons of Grace: A History of the Catholic Archdiocese of Detroit, by Leslie Woodcock Tentler, 1990

The Pottery of John Foster: Form and Meaning, by Gordon and Elizabeth Orear, 1990

The Diary of Bishop Frederic Baraga: First Bishop of Marquette, Michigan, edited by Regis M. Walling and Rev. N. Daniel Rupp, 1990

Walnut Pickles and Watermelon Cake: A Century of Michigan Cooking, by Larry B. Massie and Priscilla Massie, 1990

The Making of Michigan, 1820–1860: A Pioneer Anthology, edited by Justin L. Kesten-baum, 1990

America's Favorite Homes: A Guide to Popular Early Twentieth-Century Homes, by Robert Schweitzer and Michael W. R. Davis, 1990

Beyond the Model T: The Other Ventures of Henry Ford, by Ford R. Bryan, 1990

Life after the Line, by Josie Kearns, 1990

Michigan Lumbertowns: Lumbermen and Laborers in Saginaw, Bay City, and Muskegon, 1870–1905, by Jeremy W. Kilar, 1990

Detroit Kids Catalog: The Hometown Tourist, by Ellyce Field, 1990

Waiting for the News, by Leo Litwak, 1990 (reprint)

Detroit Perspectives, edited by Wilma Wood Henrickson, 1991

Life on the Great Lakes: A Wheelsman's Story, by Fred W. Dutton, edited by William Donohue Ellis, 1991

Copper Country Journal: The Diary of Schoolmaster Henry Hobart, 1863–1864, by Henry Hobart, edited by Philip P. Mason, 1991

John Jacob Astor: Business and Finance in the Early Republic, by John Denis Haeger, 1991

Survival and Regeneration: Detroit's American Indian Community, by Edmund J. Danziger, Jr., 1991

Steamboats and Sailors of the Great Lakes, by Mark L. Thompson, 1991

Cobb Would Have Caught It: The Golden Age of Baseball in Detroit, by Richard Bak, 1991

Michigan in Literature, by Clarence Andrews, 1992

Under the Influence of Water: Poems, Essays, and Stories, by Michael Delp, 1992

The Country Kitchen, by Della T. Lutes, 1992 (reprint)

The Making of a Mining District: Keweenaw Native Copper 1500–1870, by David J. Krause, 1992

Kids Catalog of Michigan Adventures, by Ellyce Field, 1993

Henry's Lieutenants, by Ford R. Bryan, 1993

Historic Highway Bridges of Michigan, by Charles K. Hyde, 1993

Lake Erie and Lake St. Clair Handbook, by Stanley J. Bolsenga and Charles E. Herndendorf, 1993

Queen of the Lakes, by Mark Thompson, 1994

Iron Fleet: The Great Lakes in World War II, by George J. Joachim, 1994

Turkey Stearnes and the Detroit Stars: The Negro Leagues in Detroit, 1919–1933, by Richard Bak, 1994

Pontiac and the Indian Uprising, by Howard H. Peckham, 1994 (reprint)

Charting the Inland Seas: A History of the U.S. Lake Survey, by Arthur M. Woodford, 1994 (reprint)

Ojibwa Narratives of Charles and Charlotte Kawbawgam and Jacques LePique, 1893–1895. Recorded with Notes by Homer H. Kidder, edited by Arthur P. Bourgeois, 1994, co published with the Marquette County Historical Society

Strangers and Sojourners: A History of Michigan's Keweenaw Peninsula, by Arthur W. Thurner, 1994

Win Some, Lose Some: G. Mennen Williams and the New Democrats, by Helen Washburn Berthelot, 1995

Sarkis, by Gordon and Elizabeth Orear, 1995

The Northern Lights: Lighthouses of the Upper Great Lakes, by Charles K. Hyde, 1995 (reprint)

Kids Catalog of Michigan Adventures, second edition, by Ellyce Field, 1995

Rumrunning and the Roaring Twenties: Prohibition on the Michigan-Ontario Waterway, by Philip P. Mason, 1995

In the Wilderness with the Red Indians, by E. R. Baierlein, translated by Anita Z. Boldt, edited by Harold W. Moll, 1996

Elmwood Endures: History of a Detroit Cemetery, by Michael Franck, 1996

Master of Precision: Henry M. Leland, by Mrs. Wilfred C. Leland with Minnie Dubbs Millbrook, 1996 (reprint)

Haul-Out: New and Selected Poems, by Stephen Tudor, 1996

Kids Catalog of Michigan Adventures, third edition, by Ellyce Field, 1997

Beyond the Model T: The Other Ventures of Henry Ford, revised edition, by Ford R. Bryan, 1997

Young Henry Ford: A Picture History of the First Forty Years, by Sidney Olson, 1997 (reprint)

The Coast of Nowhere: Meditations on Rivers, Lakes and Streams, by Michael Delp, 1997

From Saginaw Valley to Tin Pan Alley: Saginaw's Contribution to American Popular Music, 1890–1955, by R. Grant Smith, 1998

The Long Winter Ends, by Newton G. Thomas, 1998 (reprint)

Bridging the River of Hatred: The Pioneering Efforts of Detroit Police Commissioner George Edwards, 1962–1963, by Mary M. Stolberg, 1998

Toast of the Town: The Life and Times of Sunnie Wilson, by Sunnie Wilson with John Cohassey, 1998

These Men Have Seen Hard Service: The First Michigan Sharpshooters in the Civil War, by Raymond J. Herek, 1998

A Place for Summer: One Hundred Years at Michigan and Trumbull, by Richard Bak, 1998

Early Midwestern Travel Narratives: An Annotated Bibliography, 1634–1850, by Robert R. Hubach, 1998 (reprint)

All-American Anarchist: Joseph A. Labadie and the Labor Movement, by Carlotta R. Anderson, 1998

Michigan in the Novel, 1816–1996: An Annotated Bibliography, by Robert Beasecker, 1998

"Time by Moments Steals Away": The 1848 Journal of Ruth Douglass, by Robert L. Root, Jr., 1998

The Detroit Tigers: A Pictorial Celebration of the Greatest Players and Moments in Tigers' History, updated edition, by William M. Anderson, 1999

Father Abraham's Children: Michigan Episodes in the Civil War, by Frank B. Woodford, 1999 (reprint)

Letter from Washington, 1863–1865, by Lois Bryan Adams, edited and with an introduction by Evelyn Leasher, 1999

Wonderful Power: The Story of Ancient Copper Working in the Lake Superior Basin, by Susan R. Martin, 1999

A Sailor's Logbook: A Season aboard Great Lakes Freighters, by Mark L. Thompson, 1999

Huron: The Seasons of a Great Lake, by Napier Shelton, 1999

Tin Stackers: The History of the Pittsburgh Steamship Company, by Al Miller, 1999

Art in Detroit Public Places, revised edition, text by Dennis Nawrocki, photographs by David Clements, 1999

Brewed in Detroit: Breweries and Beers Since 1830, by Peter H. Blum, 1999

Enterprising Images: The Goodridge Brothers, African American Photographers, 1847–1922, by John Vincent Jezierski, 2000

Detroit Kids Catalog II: The Hometown Tourist, by Ellyce Field, 2000

The Sandstone Architecture of the Lake Superior Region, by Kathryn Bishop Eckert, 2000